Zoophysiology   *Volume 29*

Editors
S. D. Bradshaw   W. Burggren
H. C. Heller   S. Ishii   H. Langer
G. Neuweiler   D. J. Randall

Zoophysiology

Volumes already published in the series:

Volume 1: *P. J. Bentley*
Endocrines and Osmoregulation
(1971)

Volume 2: *L. Irving*
Arctic Life of Birds and Mammals
Including Man (1972)

Volume 3: *A. E. Needham*
The Significance of Zoochromes (1974)

Volume 4/5: *A. C. Neville*
Biology of the Arthropod Cuticle
(1975)

Volume 6: *K. Schmidt-Koenig*
Migration and Homing in Animals
(1975)

Volume 7: *E. Curio*
The Ethology of Predation (1976)

Volume 8: *W. Leuthold*
African Ungulates (1977)

Volume 9: *E. B. Edney*
Water Balance in Land Arthropods
(1977)

Volume 10: *H.-U. Thiele*
Carabid Beetles in Their Environments
(1977)

Volume 11: *M. H. A. Keenleyside*
Diversity and Adaptation in
Fish Behaviour (1979)

Volume 12: *E. Skadhauge*
Osmoregulation in Birds (1981)

Volume 13: *S. Nilsson*
Autonomic Nerve Function in the
Vertebrates (1983)

Volume 14: *A. D. Hasler*
Olfactory Imprinting and Homing
in Salmon (1983)

Volume 15: *T. Mann*
Spermatophores (1984)

Volume 16: *P. Bouverot*
Adaption to Altitude-Hypoxia
in Vertebrates (1985)

Volume 17: *R. J. F. Smith*
The Control of Fish Migration (1985)

Volume 18: *E. Gwinner*
Circannual Rhythms (1986)

Volume 19: *J. C. Rüegg*
Calcium in Muscle Activation (1986)

Volume 20: *J.-R. Truchot*
Comparative Aspects of Extracellular
Acid-Base Balance (1987)

Volume 21: *A. Epple* and *J. E. Brinn*
The Comparative Physiology of the
Pancreatic Islets (1987)

Volume 22: *W. H. Dantzler*
Comparative Physiology of the
Vertebrate Kidney (1988)

Volume 23: *G. L. Kooyman*
Diverse Divers (1989)

Volume 24: *S. S. Guraya*
Ovarian Follicles in Reptiles and
Birds (1989)

Volume 25: *G. D. Pollak* and
J. H. Casseday
The Neural Basis of Echolocation in
Bats (1989)

Volume 26: *G.A. Manley*
Peripheral Hearing Mechanisms in
Reptiles and Birds (1989)

Volume 27: *U. M. Norberg*
Vertebrate Flight (1990)

Volume 28: *M. Nikinmaa*
Vertebrate Red Blood Cells (1990)

Volume 29: *B. Kramer*
Electrocommunication in Teleost
Fishes (1990)

B. Kramer

Electrocommunication in Teleost Fishes

Behavior and Experiments

With 140 Figures

Springer-Verlag Berlin Heidelberg NewYork
London Paris Tokyo Hong Kong Barcelona

Professor Dr. BERND KRAMER
Zoological Institute
of the University
8400 Regensburg, FRG

QL
639
.1
.K74
1990

ISBN 3-540-51927-0 Springer-Verlag Berlin Heidelberg New York
ISBN 0-387-51927-0 Springer-Verlag New York Berlin Heidelberg

Library of Congress Cataloging-in-Publication Data. Kramer, Bernd. Electrocommunication in teleost fishes: behavior and experiments/B. Kramer. p. cm. – (Zoophysiol; v. 29) Includes bibliographical references and indexes.
ISBN 3-540-51927-0 (Springer-Verlag Berlin Heidelberg New York: alk. paper). –
ISBN 0-387-51927-0 (Springer-Verlag New York Berlin Heidelberg: alk. paper)
1. Fishes – Physiology. 2. Animal communication. 3. Electroreceptors. 4. Fishes – Behavior. I. Title. II. Series. QL639.1.K74 1990 597'.50459–dc20

This work is subject to copyright. All rights are reserved, whether the whole or part of the material is concerned, specifically the rights of translation, reprinting, re-use of illustrations, recitation, broadcasting, reproduction on microfilms or in other ways, and storage in data banks. Duplication of this publication or parts thereof is only permitted under the provisions of the German Copyright Law of September 9, 1965, in its current version, and a copyright fee must alway be paid. Violations fall under the prosecution act of the German Copyright Law.

© Springer-Verlag Berlin Heidelberg 1990
Printed in Germany

The use of registered names, trademarks, etc. in this publication does not imply, even in the absence of a specific statement, that such names are exempt from the relevant protective laws and regulations and therefore free for general use.

Typesetting, printing and binding: Brühlsche Universitätsdruckerei, Giessen
2131/3145-543210 – Printed on acid-free paper

Fisches Nachtgesang
A fish's nightsong

$$\begin{array}{c}
-\\
\smile\ \smile\\
-\ -\ -\\
\smile\ \smile\ \smile\ \smile\\
-\ -\ -\\
\smile\ \smile\ \smile\ \smile\\
-\ -\ -\\
\smile\ \smile\ \smile\ \smile\\
-\ -\ -\\
\smile\ \smile\\
-
\end{array}$$

Fisches Nachtgesang: Das tiefste deutsche Gedicht
A fish's nightsong: The deepest German poem

Christian Morgenstern, Gesammelte Werke: Galgenlieder. Piper, München, Zürich 1965

Preface

The aim of writing a book is to collect the currently available data on a chosen matter in order to realize a synthetic work. In our days, however, scientific methods evolve and develop at such a fast pace that because of the amount of new data appearing during the 1 to 2 years necessary for its preparation, the book may become obsolete before its publication. Hence, much courage is necessary to make such an effort and to write a book mainly presenting the state of actual knowledge.

Electric fish biology has only a short history although the striking power of strong electric fish was already known by the Ancient Egyptians. It began in 1951 with Lissmann's discovery of the weak electric signals emitted by African tropical fish. During this 40-year period, despite a large number of scientific and popularized publications, only a small number of reviews and books have appeared on electric fish biology. Most of these dealt with anatomical, physiological and biophysical aspects of the electric organs and with problems concerning the peripheral and central nervous mechanisms of electrical signal emission and reception. They seldom, or only partially, touched on behavioural or ethological aspects. It is therefore my pleasure to welcome the present book on *Electrocommunication in teleost fishes*.

Bernd Kramer's monograph, subtitled *A fish's nightsong: the deepest German poem* (C. Morgenstern) treats a particular sensory modality, the electric sense, used for communication by certain families of a class of vertebrates living in a conductive medium. It contains four chapters in which the different aspects of electrocommunication in relation to the different genera and species are discussed in a very systematic manner. In the first chapter a concise index of taxonomy and biogeography of (gymnotiform, mormyriform and siluriform) electroreceptive fish families is proposed, a useful key for comparative ethological research especially for gymnotiforms, the taxonomy of which means still a real burden for those working in this field. In the second and third chapters the author describes the characteristics (forms and patterns) of the emitted electrical signals, the structure of the specific sense organs and explains the modes of electric signal reception, the integration mechanisms of electrosensory information, as well as the input-output links (electroreception – electric emission and its neuronal command system). In the fourth chapter, the bulk of this monograph, the principles of communicating with electric signalling are systematically expounded and a comprehensive review is presented of many experimental data concerning the specific electric organ discharge patterns occurring during non-reproductive social behaviour or courtship and spawning. A large part of this chapter is devoted to an analysis of the electrocommunicating system by experimental manipulation in order to understand the meaning of the electric signal patterns emitted in response to stereotyped exogenous electric stimuli in a given experimental context.

I like this book because its encyclopedic character provides a very complete and clear picture of the current state of our knowledge on the subject. This well conceived book has the advantage over a multiauthor publication in that it covers this subject uniformly. The systematic presentation of neuroethological aspects from a comparative point of view will be useful in further investigations. It is a real pleasure to read the many minute historical notes distributed throughout the chapters evoking the time of pioneer scientists who worked in this particular field of sensory physiology. The nicely shown contrast between old and modern-day methods traces and explains the rapid widening of our understanding of neuroethological problems in electrocommunication.

I sign this preface in friendship to Bernd Kramer and in remembrance of many pleasant hours spent together in our laboratories in Gif and in Regensburg.

Gif-sur-Yvette, March 1990   T. SZABO

Acknowledgement

To write a book while engaged in teaching, research and administration at a university is an extremely difficult job. First of all, I would like to thank my family for their understanding.

Thomas Szabo stimulated my interest in communication in electric fish and gave me the opportunity to work in his laboratory for three most rewarding years. I wish to thank him and all colleagues from the Collège and Gif laboratory for their hospitality and support; last, but not least, also Dr. Helge Szabo. I am grateful for Hubert Markl's unstinting support of this research during my stay in his group at the University of Konstanz, and for an excellent and stimulating research environment. Hubert Markl also suggested that this book be written.

I wish to thank all colleagues who collaborated with or joined our group for their precious contributions and their enthusiasm. It is especially for the younger among them that this book was written. D. Burkhardt provided the photograph of an Egyptian mural showing a mormyrid, represented on the book cover. B. Otto's help with the figures is highly appreciated.

The Deutsche Forschungsgemeinschaft supported this research from the very beginning; without this invaluable support neither this book nor the work it is based on could have been achieved.

I wish to express my gratitude to H. Langer for the interest he took in this book as an editor, and his constant support. I am also grateful to Dr. D. Czeschlik for his patience with an author always late, and the excellent support provided by his team at Springer-Verlag.

Regensburg, May 1990   B. KRAMER

Contents

Introduction 1

Chapter 1. *Taxonomy of Electroreceptive Teleosts* 3

1.1　Osteoglossiformes 6
1.2　Mormyriformes 7
1.3　Gymnotiformes 10
1.4　Siluriformes 15

Chapter 2. *Electric Sensori-Motor System* 17

2.1　Electroreception in Evolutionary Perspective 17
2.1.1　Cranial Nerves and Somatic Distribution of Electroreceptors in Teleosts 19
2.1.2　Structure of Electroreceptors in Teleosts 23
2.1.3　Modes of Encoding Electrical Stimuli 28
2.1.4　Central Projections of Electroreceptive Afferents (Mormyridae, Gymnotiformes) 33
2.2　Electric Organs 38
2.2.1　Structure and Function of Electric Organs 41
2.2.2　Neural Control of Electric Organs 47
2.3　The Electric System in the Aquatic Environment 51
2.3.1　Active Electrolocation 52
2.3.2　Electrocommunication: Spatial Aspects of Sending and Receiving Electric Organ Discharges 55

Chapter 3. *Species Diversity of Electric Organ Discharge Activity* . . 59

3.1　Waveforms of Electric Organ Discharges 59
3.1.1　Mormyriformes 62
3.1.1.1　Mormyriform Pulse Species (Mormyridae) 62
3.1.1.2　Mormyriform Wave Species (*Gymnarchus*) 69
3.1.2　Gymnotiformes 70
3.1.2.1　Gymnotiform Pulse Species 70
3.1.2.2　Gymnotiform Wave Species 75
3.1.3　Siluriformes 84

3.2	Patterns of Spontaneous Discharge Rates	85
3.2.1	Mormyriformes	86
3.2.1.1	Mormyriform Pulse Species (Mormyridae)	86
3.2.1.2	Mormyriform Wave Species (*Gymnarchus*)	97
3.2.2	Gymnotiformes	97
3.2.2.1	Gymnotiform Pulse Species	97
3.2.2.2	Gymnotiform Wave Species	100
3.2.3	Siluriformes	103
3.3	Responses to Disturbances (or Food Stimuli)	104
3.3.1	Mormyriformes	104
3.3.1.1	Mormyriform Pulse Species (Mormyridae)	104
3.3.1.2	Mormyriform Wave Species (*Gymnarchus*)	106
3.3.2	Gymnotiformes	106
3.3.2.1	Gymnotiform Pulse Species	106
3.3.2.2	Gymnotiform Wave Species	107
3.3.3	Siluriformes	108

Chapter 4. *Communicating with Electric Organ Discharges* 111

4.1	Electrical and Motor Displays of Communicating Fish	112
4.1.1	Mormyriformes	112
4.1.1.1	Mormyriform Pulse Species (Mormyridae)	112
4.1.1.2	Mormyriform Wave Species (*Gymnarchus*)	131
4.1.2	Gymnotiformes	132
4.1.2.1	Gymnotiform Pulse Species	132
4.1.2.2	Gymnotiform Wave Species	135
4.2	Experimental Manipulation of the Electrocommunication System	137
4.2.1	Mormyriformes	137
4.2.1.1	Ethological Approach	137
4.2.1.2	Playback of EOD Interval Patterns	146
4.2.1.3	Species Recognition by EOD Interval Pattern (B. Kramer and H. Lücker)	157
4.2.1.4	Discrimination of Inter-Pulse Intervals (B. Kramer and U. Heinrich)	170
4.2.2	Gymnotiformes	178
4.2.2.1	Electrical Stimulation and Playback of EOD Patterns in Pulse Species	178
4.2.2.2	Electrical Stimulation and Playback of EODs in Wave Species	184

Conclusion 217

References 219

Systematic Index 235

Subject Index 239

Introduction

The unusual capacity of some tropical freshwater fishes (of the dominating subgroup *Teleostei*) to generate and sense electric signals, the discharges of their weak electric organs, was discovered by Hans Lissmann (1951, 1958) of the University of Cambridge. He demonstrated the function of an *active electrolocation system*, but, along with others, also proposed a second function, that of *communication*. Studies in electrical communication were pioneered by Patricia Black-Cleworth (1970), then in the laboratory of T. Bullock at the University of California in La Jolla, and Peter Moller (1970), then in the laboratory of T. Szabo at the CNRS research institute and the Collège de France in Paris.

Although we humans of the twentieth century are used to communicating by means of the electric channel (telephone, radio, etc.), we are unable to sense weak electric organ discharges, and cannot imagine what it feels like to communicate by this modality. Unlike the elephant-nose fishes of Africa (Mormyriformes) and the knife fishes (Gymnotiformes) of South America, we do not live right inside the conductive medium, and the electric currents associated with signalling and communication do not normally flow through our bodies.

The first two chapters of this book on electrocommunication are intended to give background information. The short first chapter is on the taxonomy of electroreceptive teleost fishes and their biogeography. The longer second chapter deals with the structure and function of the electroreceptive and the electrogenic parts of the "electric system", that is, its sensory and its motor part (comparable to a bat's SONAR system). Chapter 2 tries to integrate the major reviews by Bennett (1971a,b), Szabo (1974), Szabo and Fessard (1974), Bullock (1982), some of the reviews collated in Bullock and Heiligenberg (1986), and many other sources.

The main emphasis of the book is on the comparative study of electric signalling and communication from a behavioral perspective (Chaps. 3 and 4). The author feels that enough knowledge has been accumulated to make such an endeavor both possible and necessary.

Why write a book on behavior (even if the animals studied are highly unusual) when neurobiology appears to be going largely in a molecular direction? Behavioral studies are among the best, and sometimes the only, means we have to identify the function of a structure or system. Sometimes from behavior we know a sensory capacity before it has been possible to find the associated sensory system that must exist [for example, orientation using the earth's magnetic field in birds (review Keeton 1979; volume edited by Kirschvink et al. 1985) or fish (review Smith 1985; Walker 1984)]. Very often the sensory capacities of an animal are too complex to be deduced from the study of single neurons or small groups of such. Behavioral methods are useful to study the overall performance and the properties of such sensory systems, and may aid and guide neurophysiological analysis. The ecological significance of an adaptation cannot be approached by restrictive or destructive experimentation like, say, brain-slice techniques, but is, very often, amenable to behavioral research.

Let us take a look at these highly specialized vertebrates that have remarkable brains, both in size and complexity, not only by fish standards. These predominantly nocturnal creatures are the products of one of the least explored, remote and dangerous biotopes, tropical freshwater that is a true laboratory of evolution. These fishes offer excellent and, in certain areas, unique opportunities for the study of advanced and highly specialized sensory, motor, and communication systems, their adaptive advantages, and the selective pressures that shaped them in the course of evolution.

Chapter 1

Taxonomy of Electroreceptive Teleosts

Taxonomy is viewed by some as a dull and boring subject. This is understandable when one thinks of some old-fashioned museums that present endless shelves of, say, dead fish preserved in alcohol. Very often these fish are unattractively bleached, unnaturally oriented, difficult to see through the jar, and covered partly by an identifying tag.

In reality, taxonomy is both exciting and extremely useful. It is exciting because of all biological disciplines, it is taxonomy that confronts real life, with its endless variations of structure and function to achieve the most diverse goals. Modern taxonomy is also useful, for it tries, and largely appears to succeed, in classifying organisms in a natural, phylogenetic system that reflects the path of evolution, and finds out the degrees of relationship both within and between groups.

It is necessary to classify, given the sheer size of the job. Estimates for the number of living organisms are at least 4 to 5 million and run as high as 10 million (even 30 million; Fittkau 1985). From Linnaeus' time in the 1750s, when he introduced binomial nomenclature that is still in use, no more than a third has been described and named. Estimates of the number of living vertebrates are 45,000 (Raven and Johnson 1986). About half of these are teleosts, the most successful vertebrates in terms of species number (23,000, Lauder and Liem 1983; 30,000 according to Starck 1978). While birds and mammals have become the dominating land vertebrates, teleosts are the dominating aquatic vertebrates, ranging in size from a centimeter long when mature to giants more than 6 m long (Keeton 1980).

Whatever his field of interest or specialization, a biologist needs to know the systematic place of the organism he studies, and how it is interrelated with similar organisms. This is especially necessary for comparative research, something nearly every student of fish behavior, sensory physiology, neurophysiology, or neuroanatomy, etc., is doing.

The predictive power of a good phylogenetic system is extraordinary. By recognizing a salient feature of an unknown fish (for example, a mormyrid) that allows one to place it in the correct family (or even genus), we immediately know a lot about its biology without ever having studied that species.

Central to taxonomy is the modern species concept that has a rather convoluted history. For an introduction, the reader may refer to a general biology textbook like that of Keeton (1980), Raven and Johnson (1986), or Purves and Orians (1987).

Just as a set of anatomical, physiological, or behavioral characters defines a species, a geographical area or range is also typical of every species and necessary for a complete description. Fossil fresh water organisms, especially lungfishes, reveal a lot about the history of the earth's continents (Alfred Wegener's continental drift theory; see general biology or geology textbooks).

Osteichthyes (bony fishes in the broad sense) probably originated in fresh water and are derived from Silurian Placoderms that have been extinct for at least 230 million years (Silurian, 430-395 million years ago). Osteichthyes have become the dominant vertebrates in both fresh water and in the oceans since the Devonian, the so-called Age of Fishes (395-345 million years ago). Although known from the early Mesozoic (that is, 225 million years ago), the teleosts (bony fishes in the strict sense) have their adaptive radiation only late in the earth's history, at the close of the Cretaceous (Cretaceous, 136-65 million years ago). This was undoubtedly linked with the preceding evolution and radiation of the flowering plants (angiosperms) in the Cretaceous, which provided new ecological niches for many animals, including insects and also freshwater fishes.

Today we believe that electroreception is an ancestral vertebrate sense, still widely present in the realm of lower aquatic vertebrates, that was lost with the advent of the ancestors of teleosts and a few of their less well-known relatives (see Chap. 2 for a more detailed discussion). Electroreception in teleosts must have reevolved at least twice: in some of the African Osteoglossomorpha (bony-tongued fishes), and in the siluriform/gymnotiform assemblage of the Ostariophysi (fish with a Weberian apparatus connecting the swim bladder with the ear, giving them keen hearing). None of the remaining teleosts are known to be electroreceptive.

The Siluriformes and the Gymnotiformes are now believed to represent a monophyletic lineage (Fink and Fink 1981 even lump them together under the label Siluriformes). They probably have electroreceptors of the common (ampullary) type by common descent (see Chap. 2). However, the tuberous electroreceptors of the Gymnotiformes that are physiologically different (see Chap. 2) appear to be a specialized trait. The recent discovery of a tuberous electroreceptor in a South American siluriform (Andres et al. 1988), if confirmed as to its function, might extend the large number of shared traits between both groups. Siluriforms and gymnotiforms are unusual among teleosts in that both have an anterior lateral line nerve with a prominent recurrent branch that innervates the electroreceptors of the body trunk (see Chap. 2; see Fink and Fink 1981 for further similarities).

Traditionally, gymnotiforms have been viewed as highly modified characiforms (tetras, piranhas, pacus; see Fink and Fink 1981 for review). The present view of siluriforms and gymnotiforms as members of a monophyletic group that is a sister group of the characiforms has become possible by the application of Hennig's (1966) cladistic approach to phylogeny. He considers that shared traits among a related group of living organisms that have been acquired recently but are not present in all of its members (synapomorphies) are more revealing for a phylogenetic tree than those that are old and shared by all members of a group (symplesiomorphies).

The African Osteoglossomorpha also present evolutionary problems regarding electroreception and construction of a phylogenetic tree (Braford 1986; Finger et al. 1986). While *all* Mormyriformes that have been studied are

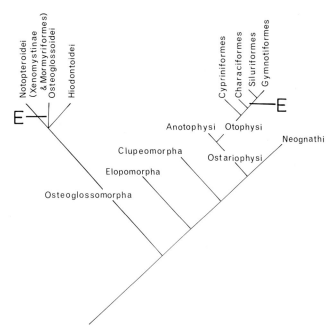

Fig. 1.1. Phylogeny of electroreceptive teleosts. The distant relationship between the siluriform/gymnotiform lineage and the notopteroids (Mormyriformes plus Xenomystinae; Lauder and Liem 1983) indicates that electroreception *(E)* must have evolved at least twice. Horizontal names are major teleost groups (modified from Finger et al. 1986, which was adapted from Lauder and Liem 1983)

electroreceptive (that is, the Mormyridae and the monotypic Gymnarchidae), this cannot be said of the related Osteoglossiformes; perhaps the Osteoglossiformes are not the monophyletic group they were believed to be. One osteoglossiform family, the Notopteridae, has two African species, *Xenomystus nigri* and *Papyrocranus (Notopterus) afer*, that are both electroreceptive, while the Asian notopterids are not. The electroreceptors of *Xenomystus* are of the common, ampullary type (preliminary data, see Braford 1986). All other Osteoglossiformes seem to lack electroreception (that is, the Osteoglossidae, Pantodontidae, and Hiodontidae, plus the Asian Notopteridae; see Fig. 1.1).

Parsimony would suggest that the African Notopteridae (the Xenomystinae) and the Mormyriformes are a monophyletic lineage (called Notopteroidei in Fig. 1.1; Lauder and Liem 1983). However, this is difficult to reconcile with the present systematics that places the Notopteridae apart from the Mormyriformes. Braford (1986) discusses alternative paths of phylogeny.

1.1 Osteoglossiformes

African and Asian notopterids superficially resemble the South American knife fishes, or gymnotiforms, which are only very distantly related. Both African notopterids are electroreceptive, *Xenomystus nigri* and *Papyrocranus (Notopterus) afer*, and are sometimes also called knife fishes. These fish have a very long anal fin fused with the tail fin, no (or rudimentary) ventral fins, a laterally compressed body, and small scales on body and head (Lévêque and Paugy 1984). They are not known to possess an electric organ and occur in fresh water only.

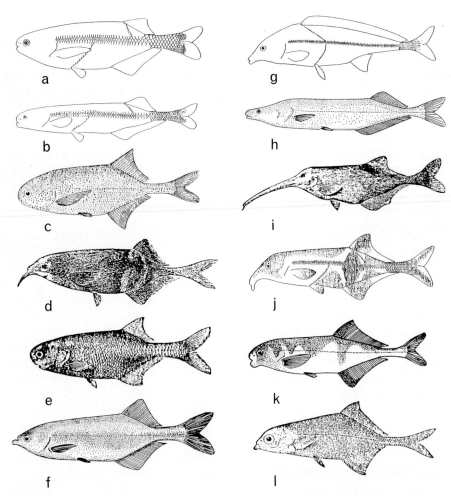

Fig. 1.2. Several mormyrids. *a Brienomyrus niger* (Günther 1866). Gambia, Niger, White Nile, Chad, Volta, and Senegal basins. Total length up to 130 mm; *b Brienomyrus brachyistius* (Gill 1863). Coastal drainages from Gambia to Zaire; *c Pollimyrus isidori* (Valenciennes 1846). Nile, Gambia, Middle Niger, Volta, and Chad basins, coastal rivers of Ivory Coast (Sassandra, Bandama, Mé). Senegal (Lévêque and

1.2 Mormyriformes

Mormyridae (Fig. 1.2). The mormyrids occur only in fresh water of Africa (predominantly fully tropical, excepting the northern Maghreb and the southern Cape regions). Of the many fish families endemic to Africa, the mormyrids are the largest: 18 genera and 188 species (subspecies excluded; Gosse 1984), about the number of living Primates (approximately 190 species; Raven and Johnson 1986), or the number of fish species in the whole of Europe (192; review Lowe-McConnell 1987). Mormyrids range in maximum size from about 9 cm when mature to 1.5 m or more. Although tremendously varied, they have a more or less "normal" fish silhouette with a full set of fins (a pair of both pectoral and ventral fins, a dorsal, an anal, and a tail fin). The tail fin is the main source of thrust; it is connected to the body by a caudal peduncle containing Gemminger's bones. Trunk muscles move the tail fin via strong tendons traversing the caudal peduncle that contains the "weak" electric organ discharging in pulse-like fashion. Mormyrids are unrivalled among fish for their huge brain, and among vertebrates for their gigantic cerebellum; some of them have bizarre trumpets or snouts that indicate specialized feeding habits. Mormyrids are principally fluviatile, with only a few species occurring as well in one of Africa's great lakes. Of these species most show potamodromous behavior in moving up affluent rivers to spawn (review Lowe-McConnell 1987), although very few details, if any, are known. Most mormyrids do not seem to like the greater salinity of lakes. An especially characteristic habitat of certain species seems to be river rapids (Roberts and Stewart 1976).

The systematics has been revised by Taverne (1972). However, some of the salient features (anatomical and osteological) are difficult to use in the field or in live fish, and the behavioral scientist finds himself more often than not in the dark as to the species identity of his object of study.

Paugy 1984). Up to 90 mm standard length; *d Gnathonemus petersii* (Günther 1862). West African rivers, from Niger to Zaire basins. Maximum total length, beyond 20 and probably below 30 cm (Kramer, pers. observ.); *e Petrocephalus bovei* (Valenciennes 1846). Nile, Chad, Niger, Volta, Gambia, and Senegal basins. Coastal drainages of Ivory Coast (Lévêque and Paugy 1984). Up to 90 mm standard length; *f Marcusenius cyprinoides* (Linné 1758). Nile, from the Delta to Bahr-el-Jebel, Niger and Chad basins. Up to 330 mm standard length; *g Mormyrus rume* (Valenciennes 1864). Gambia, Senegal, Niger, Volta, and Chad basins. Up to 870 mm standard length. Also found in several coastal basins, especially of Ivory Coast; up to 900 mm standard length (Lévêque and Paugy 1984); *h Mormyrops deliciosus* (Leach 1818). Senegal, Gambia, Niger, Chad, Zaire, Volta, and Zambezi basins, Lakes Malawi and Tanganyika, Webi Shebeli and Juba. Up to 1.50 m total length; *i Campylomormyrus numenius* (Boulenger 1898). Zaire basin, only in big rivers. Up to 650 mm total length; *j Campylomormyrus tamandua* (Günther 1864). Volta, Niger, Chad, Shari and Zaire basins. Up to 430 mm standard length; *k Hippopotamyrus harringtoni* (Boulenger 1905). White Nile, Niger river, Chad basin. Up to 305 mm standard length; *l Marcusenius greshoffi* (Schilthuis 1891). Zaire basin (lower and central). Total length 108 mm (Boulenger 1909). If not stated otherwise, nomenclature, length and distribution data according to Gosse (1984). *a,b,c,g,j* reproduced from Lévêque and Paugy (1984); *d,e,i,l* from Boulenger (1909); *f,h,k* from Daget and Durand (1981)

Table 1.1. Classification of the Mormyriformes (after Gosse 1984; references in Daget et al. 1986)

Mormyridae (188 spp.)
    Boulengeromyrus Taverne and Géry 1968, 1 sp.
    Brienomyrus Taverne 1971, 8 spp.
    Campylomormyrus Bleeker 1874, 14 spp.
    Genyomyrus Boulenger 1898, 1 sp.
    Gnathonemus Gill 1862, 5 spp.
    Heteromormyrus Steindachner 1866, 1 sp. (only one specimen known)
    Hippopotamyrus Pappenheim 1906, 16 spp.
    Hyperopisus Gill 1862, 1 sp.
    Isichthys Gill 1863, 1 sp.
    Ivindomyrus Taverne and Géry 1975, 1 sp.
    Marcusenius Gill 1862, 37 spp.
    Mormyrops Müller 1843, 26 spp.
    Mormyrus Linné 1758, 20 spp.
    Myomyrus Boulenger 1898, 3 spp.
    Paramormyrops Taverne, Thys van den Audenaerde and Heymer 1977, 2 spp.
    Petrocephalus Marcusen 1854, 20 spp.
    Pollimyrus Taverne 1971, 19 spp.
    Stomatorhinus Boulenger 1898, 12 spp.
Gymnarchidae (1 sp.)
    Gymnarchus niloticus Cuvier 1829

For a complete catalogue and a monumental bibliography of the Mormyriformes (and other African freshwater fishes) see Gosse (1984) in Daget et al. (1984, 1986); see also Table 1.1. Lévêque and Paugy (1984) provide a very useful phenetic key for part of West Africa, and Durand and Lévêque (1981) for the more global "sudanian" fish fauna (see below). A more complete key is in preparation (part of the CLOFFA-project; Daget et al., starting in 1984). A useful short introduction is given by Hopkins (1986a; along with a detailed bibliography).

Ichthyological provinces of Africa are based largely on the present drainage systems (Fig. 1.3; Lowe-McConnell 1987). There are also some inland drainage areas, such as Lake Chad or Lake Turkana to the East of the Nile. The equatorial Zaire basin has the richest fish fauna of all (690 species, excluding Lake Tanganyika with its highly endemic fauna). The Zaire basin has 75 mormyrids, or 18% of the fauna. The so-called sudanian region, especially West Africa, also has many mormyrids, with the smaller Cameroon and Gabon region containing elements from both Zaire and West Africa; this region is also very rich in species. The southern Zambezi region, however, lacks many families of fishes present in the neighboring Zaire region; it has only 101 fish species and also fewer kinds of mormyrids.

In main rivers, mormyrids are benthic or littoral fishes feeding mainly on aquatic insects (review Lowe-McConnell 1987). Daget and Durand (1981) report that most species are bottom-oriented micropredators of the benthic fauna.

With 18 species the Mormyridae are also relatively well represented in Lake Chad, a riverine lake of the sudanian region (Lowe-McConnell 1987).

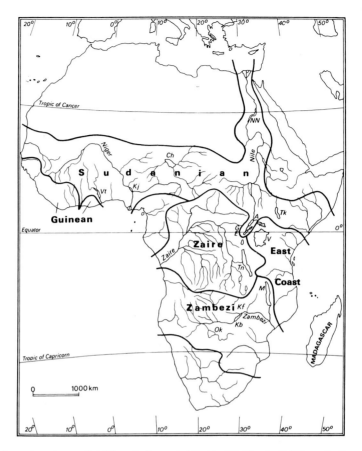

Fig. 1.3. Ichthyofaunal regions of Africa. Lakes: *A* Albert, *Ch* Chad, *E* Edward, *G* George, *Kb* Kariba, *Kf* Kafue Flats, *Kj* Kainji, *M* Malawi, *NN* Nasser/Nubia, *Ok* Okavango, *Tk* Turkana, *Tn* Tanganyika, *V* Victoria, *Vt* Volta (Lowe-McConnell 1987)

One of these, the small *Pollimyrus isidori*, occurs in open water, feeding on zooplankton (as well as, more typical for mormyrids, on insects), thus supporting top consumers like the piscivore tigerfish *Hydrocynus forskalii* (Characiformes). *P. isidori* is among that fish's main prey. (There are, however, no reports of sieve-like gill rakers in *P. isidori* that would be required for a true planktivore.) In aquaria, *P. isidori* shows bottom-oriented micro-predator behavior typical for mormyrids. Daget and Durand (1981) report that *P. isidori's* typical habitat is the inundation plains.

Lake Victoria, a huge lake the size of Switzerland, has five species of mormyrids. *Mormyrus kannume* was studied as to its food preference. It took a wide diet of chironomid, chaoborid, and trichopteran larvae, as well as ephemopteran nymphs and shrimps. There was no marked preference for any of these; a seeming preference depending on several factors and where the fish was feeding (hard or soft bottom, lunar phase, etc.) and also the size of the fish

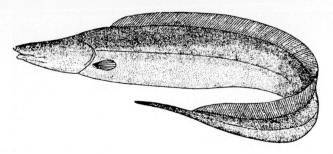

Fig. 1.4. Gymnarchus niloticus (up to 1.5 m long; Lévêque and Paugy 1984)

(see review Lowe-McConnell 1987). The main impression was that of great flexibility of feeding behavior.

Gymnarchidae. There is only one species to this family, *Gymnarchus niloticus* (Fig. 1.4). It is a large (maximum size about 1.5 m), predatory fish that is unique for its mode of locomotion with a long, undulating dorsal fin, all other fins except the pectorals being lost. *Gymnarchus* has a weak electric organ emitting a constant-frequency wave discharge; it is the only representative of this discharge type in Africa. *Gymnarchus* has been the favorite study object of Hans Lissmann for the demonstration of active electrolocation and the analysis of its mechanism (Lissmann 1958; Lissmann and Machin 1958).

The fish is widely distributed in the drainage systems of the Nile, Chad, Niger, Senegal, and is also found in Gambia and Volta.

1.3 Gymnotiformes

Gymnotiforms are freshwater fish, widely distributed through tropical South America. Although some occur as far north as Panama, their central range is the Amazon basin. Whole river systems have yet to be explored; therefore, biogeographical data and species numbers are very provisional (Lowe-McConnell 1987).

It seems that at the turn of the century the range of gymnotiforms was greater: Ellis (1913) reports gymnotiforms from as far north as the Rio Motagua in Guatemala (Central America), and as far south as the Rio de la Plata, east of the Andes; fish also occurred on the western coast of Colombia and Ecuador. Three species occurred throughout almost the entire range (*Gymnotus carapo, Eigenmannia virescens*, and *Sternopygus macrurus*); with an additional four species almost, but not quite, so widely distributed. The remaining species were largely confined to the Amazon system and the Guianas. One species, *G. carapo*, even occurred on the West Indies, although Ellis (1913; p. 157, Table) does not give further detail.

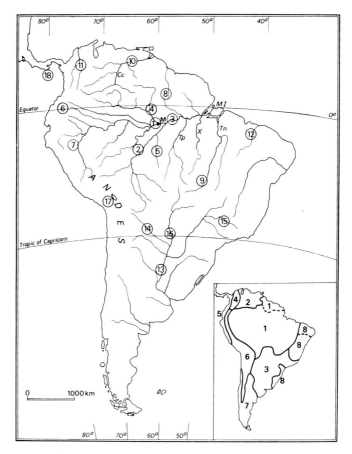

Fig. 1.5. Major river systems and sites of ichthyological studies in South America (*large map*), and ichthyofaunal provinces (*inset*). The eight provinces are (*inset*): *(1)* the Guyanan-Amazonian region, *(2)* the Orinoco-Venezuelan region, *(3)* the Paranean region, *(4)* the Magdalenean region, *(5)* the Trans-Andean region, *(6)* the Andean region, *(7)* the Patagonian region, and *(8)* the East Brazilian region. Study sites (*large map*): *1* Central Amazon at *M* Manaus; *2* Madeira; *3* Itacoatiara; *4* Negro; *5* Aripuaña; *6* Ecuador; *7* Peruvian Amazon; *8* Rupununi; *9* Mato Grosso; *10* Orinoco; *11* Magdalena; *12* Paranaìba; *13* Paranà; *14* Pilcomayo; *15* Mogi Guassu; *16* Gran Chaco; *17* Lake Titicaca; *18* Panama streams. *Cc* Casiquaire canal, connecting temporarily regions 1 and 2; *MI* Marajo Island; *Tn* Tocantins; *Tp* Tapajós; *X* Xingu (Lowe-McConnell 1987)

Lowe-McConnell (1987) distinguishes eight ichthyofaunal provinces in South America (as recognized by Géry 1969; Fig. 1.5): "(1) the Guyanan-Amazonian region with interconnections to (2) the Orinoco-Venezuelan region to the north, and (3) the Paranean to the south; (4) Magdalenean and (5) Trans-Andean in the northwest; (6) Andean and (7) Patagonian south of this, with (8) the East Brazilian in rivers flowing to the Atlantic coast".

Most information on gymnotiforms comes from the central Amazonian system, an area almost the size of the continental US, which contains the world's richest freshwater fish fauna (probably more than 1300 species, Lowe-McConnell 1987; close to 2000 species, Fittkau 1985). Colinvaux (1989) and Fittkau (1985) review hypotheses that may explain the evolution of this unique ecosystem and its (selective) species-richness. Gymnotiforms also occur in Andean streams at 340 m altitude but the considerable species richness of the ichthyofauna begins in water at less than 300 m. Even the Amazonian central basin of below 200 m above sea level is an area of 2.5 million km^2 (Lowe-McConnell 1987), or 10 times the size of the German Federal Republic, or almost one-quarter the whole of Europe. Another relatively well-studied area is the Orinoco-Venezuelan region, while considerably less, if anything, is known for the other regions.

South America is a low land compared with Africa and lacks large lakes (except seasonal, lateral várzea lakes resulting from river inundations). There are, however, trumpet-shaped river mouths of certain tributaries of the Amazon river that resemble lakes (for example, the Xingu and the Tapajós; "drowned" rivers, Sioli 1984). Conditions of vital importance for fish, such as

---

Table 1.2. Classification of the Gymnotiformes (after Mago-Leccia 1978)[a]

Sternopygidae (11 spp.)
    Sternopygus Müller and Troschel 1849
    Archolaemus Korringa 1970
    Eigenmannia Jordan and Evermann 1896
    Distocyclus Mago-Leccia 1978
    Rhabdolichops Eigenmann and Allen 1942
Rhamphichthyidae (2 spp.)
    Rhamphichthys Müller and Troschel 1849
    Gymnorhamphichthys Ellis 1912
Hypopomidae (12 spp.)
    Hypopomus Gill 1864
    Hypopygus Hoedeman 1962
    Steatogenys Boulenger 1898
    Parupygus Hoedeman 1962
Apteronotidae (Sternarchidae in the older literature; 25 spp.)
    Adontosternarchus Ellis 1912
    Apteronotus Lacépède 1800
    Sternarchogiton Eigenmann 1905
    Sternarchella Eigenmann 1905
    Porotergus Ellis 1912
    Sternarchorhynchus Castelnau 1855
    Ubidia Miles 1945
    Orthosternarchus Ellis 1913
    Sternarchorhamphus Eigenmann 1905
    Oedemognathus Myers 1936
Gymnotidae (3 spp.)
    Gymnotus Linnaeus 1758
Electrophoridae (1 sp.)
    Electrophorus electricus Gill 1864

---

[a] Provisional species numbers according to Lowe-McConnell (1987) may show relative species abundance, even if almost certainly too low in most families (except the Electrophoridae).

the course of rivers and water level, fluctuate considerably (15 m at Manaus that is itself only little more than 30 m above sea level, although 1500 km from the sea when following the river). Certain species (perhaps many?) respond to this environmental instability by performing long migrations between feeding and spawning grounds, such as the characid *Prochilodus scrofa* from the Paranà region with its 1200 km/year round trip (review Lowe-McConnell 1987). For the majority of species, and gymnotiforms in particular, we know very little or nothing about their ecology (excepting papers by Schwassmann; Hagedorn; and Lundberg and Stager 1985).

At present there is no useful key to the gymnotiforms; a precondition to correcting the situation described above. There is no key because the systematics of the Gymnotiformes is confused in several genera. The most thorough modern work of gymnotiform classification is that of Mago-Leccia (1978) whose system is used in this book (Table 1.2). (There will be a key to all known genera and a complete list of species and synonyms by Mago-Leccia, in manuscript at the time of this writing; pers. comm.). For a short overview, see Hagedorn (1986).

Sternopygidae. These are all of the wave-discharge type. Frequencies range from very low (below 20 Hz) to high (around 800 Hz). Some species are solitary, others gregarious; most are insectivorous (especially larvae of dipterans). *Rhabdolichops* species with gill rakers are reported to feed on zooplankton (Lundberg et al. 1987). *Sternopygus* is considered the most primitive gymnotiform (Fink and Fink 1981; Lundberg and Mago-Leccia 1986 seem to agree). A recent analysis of the phylogeny within the family is given by Lundberg and Mago-Leccia (1986), who also describe four new *Rhabdolichops* species; see also Schwassmann and Carvalho (1985). *Eigenmannia* (Fig. 1.6A) is *the* gymnotiform of the tropical fish trade; hence, of most behavioral and physiological/neuroanatomical studies. Unfortunately, its systematics is confused.

Rhamphichthyidae. This family includes the sandfish, *Gymnorhamphichthys*, that buries into sand during the day (Fig. 1.6B). Members of the genus *Rhamphichthys* may grow very large (longer than 1 m); although compressed and slender, they are eaten chopped in pieces by local people near Manaus. These fish have a pulse-type discharge.

Hypopomidae. Most species are rather small (some worm-like (Fig. 1.6D); others, like *Steatogenys*, resemble fallen leaves; Fig. 1.6C) and have a pulse-type discharge. A recent systematic study is that of Schwassmann (1984).

Apteronotidae. The only gymnotiforms that have an albeit tiny caudal fin, and a filamentous dorsal fin that is usually invisible because it rests in a dorsal groove. Apteronotids are the only electric fish that have neurogenic electric organs, mostly of high or very high frequencies and often complex waveshapes (Kramer et al. 1981). The systematics of the genus *Adontosternarchus* (Fig. 1.6E) has been reanalyzed recently (Mago-Leccia et al. 1985).

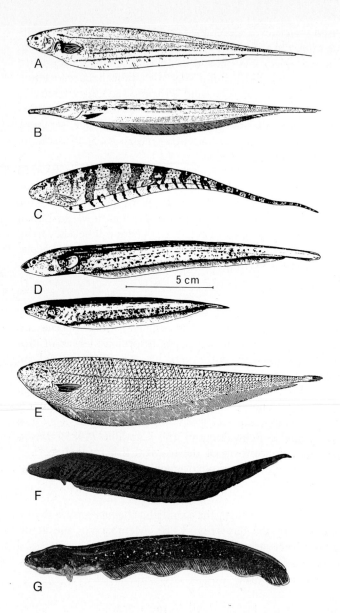

Fig. 1.6. Some Gymnotiformes. *A* Sternopygidae: *Eigenmannia* (Valenciennes). According to Mago-Leccia (pers. comm. 1990), erroneously designated *virescens* by Mago-Leccia (1978). Length, 21 cm; rio Guanare (Venezuela) (Mago-Leccia 1978); *B* Rhamphichthyidae: the sandfish *Gymnorhamphichthys hypostomus* Ellis. Length, 21.5 cm (Ellis 1913); *C* Hypopomidae: *Steatogenys elegans* (Steindachner). Up to 12.3 cm standard length (in gymnotiforms, from snout to end of anal fin). (Schwassmann 1984); *D* Male and female *Hypopomus occidentalis* from Panama (Hagedorn and Carr 1985); *E* Apteronotidae: *Adontosternarchus balaenops* (14 cm long; Ellis 1913). Name corrected according to Mago-Leccia et al. (1985); *F* Gymnotidae: *Gymnotus carapo*; *G* Electrophoridae: *Electrophorus electricus*, the electric eel, up to 2.5 m long (M. Bourgeois, from Lamarque 1979)

Gymnotidae. The voracious *Gymnotus carapo* often appears in the tropical fish trade; it is not only insectivorous like most other gymnotiforms (except, of course, the eel) but also preys on fish (Fig. 1.6F). This fish's cross-section is nearly cylindrical, not knife-like. The organ discharge is of the pulse-type.

Electrophoridae. This is a monotypic family with only one species, the strong-electric eel, *Electrophorus electricus* (Fig. 1.6G). Both strong and weak discharges are of the pulse-type.

The natives generally avoid this fish because an eel 5 ft. (about 1.5 m) long can give so powerful electric shocks that they are "sufficient to knock a man down" (Ellis 1913). Fish may reach 2.5 m and weigh some 20 kg (Lamarque 1979).

1.4 Siluriformes

The Siluriformes deserve mention, not only for the electric catfish of the monotypic family Malapteruridae with its strong organ. The distribution of this large predator is almost pan-African (from the sudanian to the Zambezi regions, present in rivers and lakes).

The Siluriformes possess ampullary (low-frequency) electroreceptors as a group. The Siluriformes are the only electroreceptive teleosts that are distributed worldwide, and are not restricted to the tropics; two families have become secondarily marine (the Ariidae, seacatfishes, and the Plotosidae, eeltail catfishes). There are about 2000 catfishes in 31 families, with 13 families endemic to South America (review Lauder and Liem 1983), and three families endemic to Africa (Lowe-McConnell 1987). According to an unpublished congress report (Hagedorn and Finger 1986), two African catfishes of the genus *Synodontis* (Mochokidae, squeakers) possess weak electric organs, like the Mormyriformes or the Gymnotiformes. A study of the function of these weak organs is planned (M. Hagedorn, pers. comm.). The systematics of African catfishes is found in Daget et al. (1984, 1986), and keys are available from Durand and Lévêque (1981) and Lévêque and Paugy (1984; Fig. 1.7). A South American catfish may have high-frequency electroreceptors (Andres et al. 1988; see Chap. 2).

Fig. 1.7. Malapterurus electricus, the electric catfish, up to 1.2 m long (Lévêque and Paugy 1984)

Chapter 2

Electric Sensori-Motor System

2.1 Electroreception in Evolutionary Perspective

The lateral line system of fishes and some amphibia is a major, complex sensory system containing several kinds of peripheral sensory organs (some of them occurring in huge numbers per individual), with their afferent and efferent connections to the brain, with specialized ganglia, nuclei or laminae within certain brain areas, as well as specific fiber tracts to higher brain centers. Cutaneous sensory organs distributed over the head and trunk (often forming a *lateral line*), such as the neuromasts or the canal organs, detect pressure waves or currents in the water; if present, electroreceptor organs form part of the system.

Following the discovery of electroreception in the Mormyriformes and Gymnotiformes by Lissmann (1951, 1958), electroreception appeared to be a truly exotic sensory modality, only found in remote, tropical freshwater biotopes of Africa and South America in a few specialized fishes living in muddy waters (however, many waters are clear, and most are not muddy). Electroreception in these nocturnal fishes appeared to be linked to their weak electric organ discharges, the energy source of a system for active electrolocation (and, as shown only later, communication).

The discovery of electroreception raised interest in an early observation of responses of an ictalurid catfish (which does not possess an electric organ) to weak electric fields (Parker and van Heusen 1917); the renewed interest led to the recognition that catfish possess lateral line electroreceptors of the ampullary organ type (see below; the long-known "small pit organs"; Roth 1968). The ampullae of Lorenzini of cartilaginous fishes, of which afferent nerve impulses were recorded as early as 1895 (review Murray 1974), were also regarded in light of a possible electroreceptive function after Lissmann's discovery. After about half a century of having been assigned various functions (this is a philosophically interesting case history of science; Bullock 1974) these receptors were, finally, recognized to be extremely sensitive to weak electric currents (Murray 1960), and to serve an electroreceptive role in the fishes' life (Dijkgraaf and Kalmijn, several papers in the 1960s; see reviews Kalmijn 1974, 1988).

It was only with the late 1970s that we began to recognize that electroreception, in the form of ampullary receptor organs of the lateral line system, is present in *all* jawed fishes except most Teleostei (and a few relatives, the bowfin *Amia* and the gars; that is, the Neopterygii, see Fig. 2.1). Electroreception was probably present in the earliest bony fishes, as suggested by

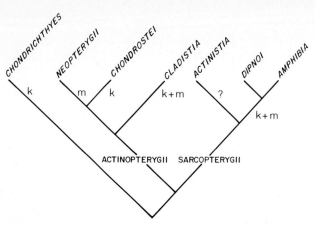

Fig. 2.1. Cladogram of the interrelationships of jawed fishes and their descendants in relationship to electroreception. *k* kinocilium present on the apical membrane of an ampullary electroreceptor cell; *m* microvilli (*?* indicates "uncertain status"). All of these taxa, with the notable exception of the Neopterygii including the teleosts, are electroreceptive, probably by inheritance from common ancestors. Only after the advent of teleosts did electroreception reevolve at least twice (in certain Osteoglossomorpha and in ostariophysan ancestors of the Gymnotiformes and Siluriformes). Besides the lack of a kinocilium, teleost ampullary receptor cells are distinguished by their sensitivity to weak anodal rather than cathodal voltages outside the receptor opening, as observed in all other taxa (Northcutt 1986)

its presence in extant representatives of all major taxa except the Neopterygii, and fossil evidence from the cosmine-layer of the dermal armor of certain lobe-finned fishes (see review Northcutt 1986). Electroreceptive nonteleosts include the Chondrostei (sturgeons and paddlefish), the Cladistia (the bichirs, *Polypterus*), the Dipnoi (the lungfishes) and even two of the three amphibian taxa (review Fritzsch and Münz 1986).

It is likely that the ancestors of the bony fishes (Osteichthyes) and their sister group, the cartilaginous fishes (Chondrichthyes: sharks, rays, chimaeras or ratfishes) were also electroreceptive (reviews Northcutt 1986; Bodznick and Boord 1986). We may even include the ancestors of the most primitive vertebrates, as suggested by the discovery of electroreceptors, termed end buds, and their central connections in the jawless lampreys (Bodznick and Northcutt 1981; review Ronan 1986).

Electroreception by common descent at least in the jawed fishes (excepting the few teleosts that are electroreceptive) is suggested by special structural and functional properties of the receptor cells: there is, as in hair cells of the mechanoreceptive lateral line system, a kinocilium at the apical membrane of a cell that faces the ampullary lumen (with an $8+1$ or $9+1$ arrangement of microtubules). The afferent nerve fiber responds to a weak stimulus voltage that is cathodal at the receptor opening. The receptor cell contacts the afferent nerve fiber by a chemical synapse with a presynaptic ribbon; there are no efferent nerve fibers terminating on the receptor cell, unlike mechanoreceptive hair cells. The receptor cell may or may not have microvilli at its apical, lumenal face.

Concerning the significance of the kinocilium, Ronan (1986) observes that its presence (rather than that of microvilli) need not represent the ancestral condition in vertebrates. The end bud receptor cells of lampreys bear microvilli but no kinocilia. Northcutt (1986), from a cladistic point of view, considers it most likely that the earliest electroreceptors in vertebrates had both kinocilia and microvilli.

Electroreceptive teleosts, on the other hand, have electroreceptor cells that lack a kinocilium but do possess microvilli; the afferent nerve fiber increases its firing rate to a weak stimulus voltage of opposite polarity (that is, anodal at the receptor opening), compared to the rest of the jawed fishes, including the jawless lampreys. In teleosts, ampullary electroreceptors (and, for that matter, also tuberous receptors) must have arisen at least twice: (1) in certain African Osteoglossomorpha, and (2) in the South American Gymnotiformes and the ubiquitous freshwater catfishes (including a marine species that has been investigated, *Plotosus*; the latter two taxa both belong to the ostariophysan subgroup of teleosts). Northcutt (1986) discusses how electroreception that was probably lost with the origin of neopterygians for unknown reasons (but see Northcutt) may have reevolved only after the origin of teleost fishes (in a way that strikingly resembles the ancestral condition in both form and function, although at a finer level of analysis many differences are also apparent; see also Braford 1986).

2.1.1 Cranial Nerves and Somatic Distribution of Electroreceptors in Teleosts

The lateral line system is served only by cranial nerves. These nerves, like those of the nose, eye, and ear, are called "special" sensory nerves because of their distinctive nature, not corresponding to the serial spinal nerves. Sensory fibers from the lateral line system are, among other fibers, found in the facial nerve (VII), the glossopharyngeal nerve (IX), and in the vagus (X), but not in the statoacoustic nerve (VIII) which serves the organs of equilibrium and hearing (see, for example, Hildebrand 1982; Fig. 2.5). However, the so-called secondary sensory cells of the statoacoustic system are hair cells that are very similar to those of the lateral line system; there is reason to believe that both are phylogenetically closely related.

Northcutt (1986) feels that the traditional view of cranial lateral line nerves should be modified because these nerves are distinct from the other cranial nerves, both in terms of embryonic origin and central termination. He argues that all lateral line organs of the head are innervated by a pair of anterior lateral line nerves, whereas all lateral line organs of the trunk and tail are innervated by a pair of posterior lateral line nerves (but see the significant modification of this pattern in the electroreceptive Gymnotiformes, Fig. 2.5, and Siluriformes; Szabo 1965; review 1974; Fink and Fink 1981). Each of these nerves possesses several ganglia; these ganglia and the organs they innervate arise from different neurogenic placodes. This suggests, according to Northcutt, that the so-called anterior and posterior lateral line nerves and the

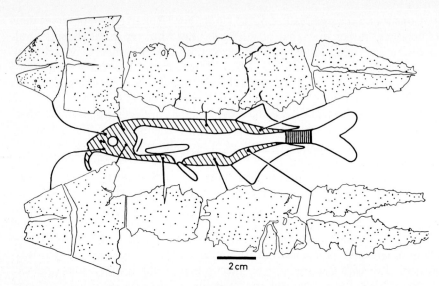

Fig. 2.2. The somatic distribution of ampullary electroreceptor organs ("small pores") in the flattened electroreceptor epidermis (*hatched*) of a *Gnathonemus petersii* (Mormyridae). *White areas* no electroreceptors present; the electric organ is located in the peduncle of the tail fin, as indicated (Harder 1968)

organs they innervate are not single phylogenetic units, but composite nerves with multiple phylogenetic origins and subsequent histories (see also Szabo 1974) that are, in spite of an extensive literature, only poorly understood.

Electroreceptors are found in four of the about 33 orders of teleost fishes (see, for example, Rosen 1982; Nelson 1984): the Mormyriformes, the Gymnotiformes, and the Siluriformes (see the detailed reviews of Bennett 1971b; Szabo 1974). A recent addition are the two African knifefishes, especially the "false featherfin" *Xenomystus nigri*, but also *Papyrocranus afer*, of the Osteoglossiformes (but no other osteoglossiforms; review Braford 1986). Mormyriformes and Osteoglossiformes are closely related (Osteoglossomorpha), while the Gymnotiformes and the Siluriformes represent a monophyletic radiation of the Ostariophysi (Fink and Fink 1981). Osteoglossomorpha and Ostariophysi are only distantly related (see Chap. 1).

The above teleosts have tonic "ampullary" receptor organs in common, while the Mormyriformes and the Gymnotiformes have in addition phasic "tuberous" receptors. Recently, however, a silurid was also found to possess a tuberous electroreceptor, but the physiology and behavior are unstudied at the time of this writing (Andres et al. 1988; see Sect. 3.2.3). The following discussion focuses on the teleosts, Gymnotiformes and Mormyriformes, possessing weakly electric organs for electrolocation and communication.

In most mormyrids, the somatic distribution of electroreceptor organs is restricted to that part of the skin which has been called the "electroreceptor epidermis" (Fig. 2.2; Harder 1968). Electroreceptors are found over the entire head and ventral and dorsal surfaces, but are absent on the sides of the body

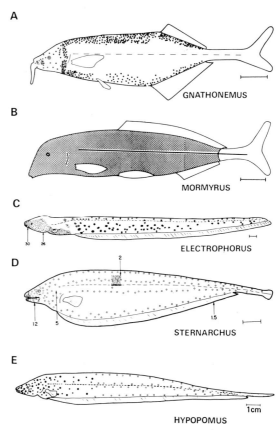

Fig. 2.3. The somatic distribution of electroreceptor organs in mormyrids and gymnotiforms. *A Gnathonemus petersii*, tuberous electroreceptors. The Knollenorgan distribution is only shown for the head while mormyromasts are shown only for the trunk. Both types extend over head and trunk, but not the sides of the body and the caudal peduncle (as shown in Fig. 2.2). *B* In *Mormyrus caballus* only the tail region and the fins lack electroreceptors. *C* The electric eel (Gymnotiformes) shows a patchy distribution of electroreceptors over most of its body. *D Diamonds* ampullary organs in *Sternarchus*. Center area of 5 mm^2 distribution of tuberous (*dots*), ampullary, and canal organs (*squares*) of the lateral line. E Tuberous organs in *Hypopomus*. Large dots pulse marker (or M) receptors. *Arrows* in *C,D*: the numbers give the density of tuberous organs/mm^2 (*A,B* from Quinet in Szabo 1974; *C–E* Szabo 1974)

and on the caudal peduncle (where the electric organ is located). *Gnathonemus petersii* is a representative of the most common distribution pattern (Fig. 2.3A) while *Mormyrus caballus* with its electroreceptors distributed over almost all of its body represents an extreme case (Fig. 2.3B).

In addition to the tonic ampullary receptors there are two kinds of phasic tuberous receptors, the mormyromasts and the Knollenorgane. The distribution of the mormyromasts (Fig. 2.3A) resembles that of the ampullary receptors (Bennett's "small pores", Fig. 2.2), but there are about twice to three times as many mormyromasts compared to ampullary organs, depending on the species. The number and distribution of the Knollenorgane (Bennett's

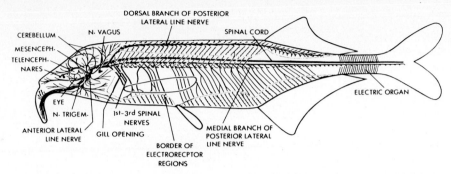

Fig. 2.4. Innervation of electroreceptors by anterior (head) and posterior (trunk) lateral line nerves in the mormyrid, *Gnathonemus petersii*. Ventrally located receptors are innervated only by the medial branch of the posterior lateral line nerve; dorsally located receptors by both the dorsal and medial branch. Mechanoreceptive canal organs are found at the level of the medial branch. Both branches carry also mechanoreceptive afferences (Harder 1968; from Bennet 1971b)

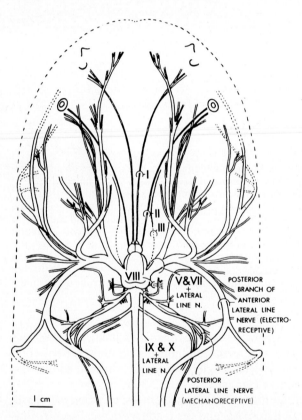

Fig. 2.5. Cranial nerves of the electric eel (Gymnotiformes). Only the anterior lateral line nerve carries electroreceptor afferences; the posterior lateral line nerve is exclusively mechanoreceptive. The anterior branch of the anterior lateral line nerve innervates cranial electroreceptors and canal organs; its posterior branch only electroreceptors from the trunk (De Oliveira Castro in Bennett 1971b)

"large pores"; Fig. 2.3A) is similar to that of the ampullary organs; however, they are less uniformly distributed with their highest density in the gill region (opercula). These electroreceptors, as well as mechanoreceptive lateral line organs, are innervated by either *posterior* or *anterior lateral line nerves*. From the posterior nerve a large dorsal branch splits off which innervates predominantly electroreceptors from the dorsal region of the trunk, while the ventrally located electroreceptors, as well as mechanoreceptive organs near the medial branch of the posterior lateral line nerve, are innervated by this medial branch (Fig. 2.4).

In the monotypic representative of the family Gymnarchidae (Mormyriformes), *Gymnarchus niloticus*, there are also ampullary and two types of tuberous receptors, gymnarchomasts type I and II (Szabo 1974). They are found everywhere in the skin, decreasing in density towards the tail. Over 40,000 gymnarchomasts type I have been estimated in a fish of 25 cm in length.

Also in gymnotiforms the density of ampullary organs is greatest on the head and decreases toward the tail (see Fig. 2.3D). The much more numerous tuberous organs show a similar trend but subtypes are restricted to certain regions differing among species (Szabo 1974). The electroreceptors of the entire body are innervated by the anterior lateral line nerve from which the large posterior branch, which carries exclusively electroreceptive fibers, splits off to join the smaller posterior lateral line nerve just behind the head (Fig. 2.5). This latter nerve is purely mechanoreceptive (neuromasts and canal organs).

The ampullary receptors of catfish tend to occur all over the body, including fins, and are sometimes arranged in orderly rows, but are more concentrated on the head and opercula. The marine catfish *Plotosus* shows a unique pattern resembling in part that of elasmobranchs as do its ampullary receptor organs with their long canals (Szabo 1974). They are clearly more sensitive to weak stimulus voltages than those of freshwater catfish (review Bennett and Obara 1986).

2.1.2 Structure of Electroreceptors in Teleosts

In freshwater electric fish, electroreceptors are visible at weak magnification of a dissecting microscope, especially in pigmented areas of the skin, because the receptors themselves are unpigmented (Fig. 2.6).

For an understanding of the electroreceptive function, the structure of the skin, as well as that of the electroreceptors and their accessory structures, is just as important as the morphology of the eye or ear for their respective functions.

The skin of freshwater electric fish (at least that of the mormyrids and the gymnotiforms) is of high resistance; the anatomical basis apparently being a region of flattened cells on top of the stratum spinosum. In sections parallel to the skin surface of mormyrids, these cells are closely packed (no intercellular

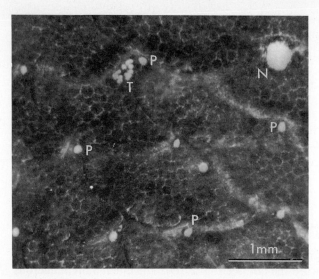

Fig. 2.6. Receptor organs in the skin of *Gymnotus* are visible as unpigmented spots. *P* phasic, tuberous, and *T* tonic, ampullary electroreceptors; *N* neuromasts (Bennett 1971b)

spaces) and regularly hexagonally arranged, resembling a botanical preparation. These cells contact each other by many desmosomes (tight junctions) and have a high content of tonofilaments (Szabo 1974). Above an electroreceptor organ this high resistance layer is punctured by a plug of "loose" tissue, that is, there is a channel across the skin where current can flow through the receptor. "In mormyrids the skin resistivity is about 50 k$\Omega \cdot$ cm^2, which is some hundred times greater than that of goldfish or hatchetfish (the latter is a South American freshwater fish). The skin resistivity is much smaller in gymnotids, 1–3 k$\Omega \cdot$ cm^2, but is still greater than in the other freshwater fish." (Bennett 1971b).

An ampullary receptor, as found in gymnotiforms and mormyriforms, is schematically shown in Fig. 2.7A. The ampulla proper lies rather deep in an epidermal invagination into the underlying corium that is surrounded by the epidermal basement membrane. The lumen of the ampulla is connected to the outside by a jelly-filled, open canal which is lined with several layers of overlapping, flattened cells. The canal wall appears sealed by the layer of flattened cells, presumably giving the skin its high resistance. The sensory epithelium at the base of the ampulla is composed of several sensory cells and a large number of accessory cells. Usually, several ampullae are served by a single afferent, branching nerve fiber.

Only a small part of the apical surface of a receptor cell is exposed to the ampullary lumen. The sensory cells of teleost ampullary organs bear microvilli (but no kinocilia).

Tuberous organs (Fig. 2.7C-F) are also located in epidermal invaginations into the corium. Unlike ampullary organs there is usually no true canal opening to the skin surface (one exception occurs in the mormyrids; Szabo 1974).

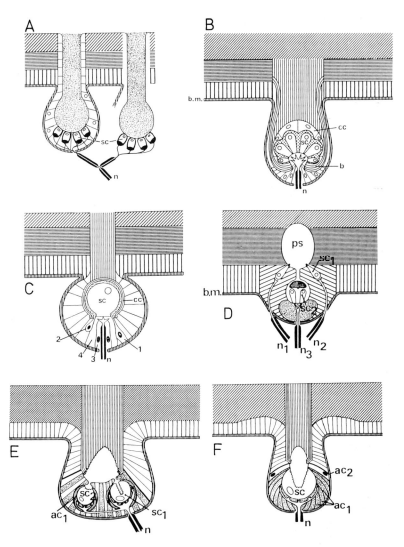

Fig. 2.7. Schematic electroreceptor organs in teleosts, located in invaginations of the epidermis. *A* Ampullary receptor; *B* tuberous receptor from a gymnotiform; *C* Knollenorgan (one kind of mormyrid tuberous receptor); *D* mormyromast (other kind of mormyrid tuberous receptor). Peculiar mucoid ball on top of sensory cells type 2. *E* Gymnarchomast type I; *F* gymnarchomast type II of *Gymnarchus* (Mormyriformes). Note stratification of epidermis in Gymnotiformes and mormyrids (but not *Gymnarchus*) by flattened cell layer with many zonular tight junctions and high tonofilament content, presumably giving the skin its high electrical resistance (*horizontally hatched*). This current barrier is punctuated above an electroreceptor, either by a jelly-filled canal (*A*), a plug of loose cells with wide intercellular spaces (*B,C*), or a mucopolysaccharide-filled perisensory space (*D*). *ac* Accessory cells; *b* capsule wall; *bm* basement membrane; *cc* covering cells; *n* nerve terminals; *ps* perisensory space; *sc* sensory cells. *Numbers* refer to different types (Szabo 1974)

Functionally, however, there is a canal across the skin, because the space covering the receptor is filled by rather loose plug cells; that is, the high resistance layer of the skin is punctured also above tuberous receptors.

In mormyrids there are two anatomically distinct types of tuberous receptors, the Knollenorgane (K) and the mormyromasts (D), which are the beginnings of two distinct sensory pathways in the brain. In gymnotids there is only one anatomical type of tuberous receptor; it can, however, apparently be subdivided into subtypes on morphological and physiological grounds.

The Knollenorgane are generally believed to serve the communication and the mormyromasts the electrolocation function of the electric sensorimotor system, although this is, in some regards, an oversimplification (see below). The segregation of functions is less obvious in gymnotiforms.

The Knollenorgane have 1-35 receptor cells, with each cell located in its own perisensory cavity within the sensory organ, while in the tuberous receptor of gymnotiforms 10 to 100 cells share one cavity. About 90% of the apical surface of a tuberous receptor cell is exposed to the perisensory space; the surface is richly decorated with microvilli. The basal part of a sensory cell rests on a hillock of supporting cells; there is only one (branching) afferent nerve fiber contacting all cells of an organ. The Knollenorgan sensory cell is capable of spiking; transmission to the postsynaptic membrane is thought to be electrotonic and fast, although in tuberous organs only chemical synapses have been described to date. [However, in the gymnotiform *Sternarchus* gap junctions in addition to chemical ribbon synapses were found (reviewed in Szabo 1974); gap junctions are known to be fast, electrotonically transmitting and sometimes to occur together with chemical synapses in the same cell.]

The tuberous organs of pulse gymnotiforms can be functionally and anatomically divided into pulse marker units (M units) and burst duration coders (B units). Anatomical details of the contact zone of the afferent nerve fiber with the sensory cells differ between these two types (see review Zakon 1988). Synapses are chemical.

The tuberous organs of wave gymnotiforms (especially Sternopygidae) are divided into phase coders (T units) and probability coders (P units) on physiological grounds; it is not certain whether anatomical correlates exist, and even the physiological distinction has been doubted (Viancour 1979a). Based on the number of receptor organs contacted by a single afferent fiber, Zakon (1988) considers that he can anatomically distinguish P from T receptors in *Sternopygus* and *Eigenmannia*.

The mormyromasts are the most numerous of all electroreceptors in the mormyrid skin: in a 13 cm *Gnathonemus petersii*, Harder (1968) counted 2296 mormyromasts, 911 ampullary and 986 Knollenorgane. The mormyromast is the most complex of all electroreceptors; a *sense organ* (Szabo). Two different types of sensory cells are innervated separately. There is an outer and an inner sensory cavity, one for each sensory cell type, which are connected by a small canal. The skin's flattened cell layer of high resistance is punctured above a mormyromast by the bigger outer sensory cavity (instead of a plug of loose tissue with large intercellular spaces, as observed in most other tuberous receptors).

The bottle-shaped sensory cells of the outer, larger sensory cavity of a mormyromast are embedded in its wall and contact the lumen only by their apical tips; they are unique among electroreceptors (and hair cells in general) by not bearing microvilli nor any other specialization in their apical cell region. Until recently, their status as electroreceptor cells appeared mainly derived from their innervation, as there are no intracellular recordings; recordings from afferent nerve fibers could not be attributed to either sensory cell type (Kramer-Feil 1976; but see below). Each of the five to seven outer sensory cells is contacted by one of two or three branching nerve fibers.

The other, inner sensory cell type is more similar to that seen in the Knollenorgan, and especially to that in gymnotiforms: 3-5 elongated cell bodies stand upright in the deeper sensory chamber, exposing most of their microvilli-covered surface to the chamber fluid. A single, branching nerve fiber contacts these cells. Still not known is the function of the unique mucopolysaccharide "ball" resting on top of the inner sensory cells, depressing their surface; the ball is free of calcium and therefore does not represent a statolith (Denizot in Szabo 1974).

Bell et al. (1989) now have shown that the afferent fibers from both cell types project to the electrosensory lobe of the lateral line (ELL), by using retrograde labeling with horseradish peroxidase. Fibers from the outer sensory cells terminate in the medial zone while fibers from the inner sensory cells terminate in the dorsolateral zone of the ELL (see Sect. 2.1.4). Also physiological data supporting an electrosensory function of both sensory cell types are now available (Bell 1990; see below).

The morphology of gymnarchid tuberous electroreceptors, gymnarchomasts, is strikingly different from that of mormyrids (Fig. 2.7), despite the close phylogenetic relationship of both families. As the physiology of gymnarchomasts has been little studied (with the exception of Bullock et al. 1975; Table 2.1), and occur in just one species, *Gymnarchus niloticus*, the only African fish with a wave discharge, gymnarchomasts are dealt with here only in passing. It may, however, be noted that all gymnarchid electroreceptor cells (including ampullary) are unique in their more or less deep, microvilli-covered

Table 2.1. Summary of physiological, tuberous receptor types for both orders of weakly electric teleosts[a] (Zakon 1988)

Gymnotiforms			Mormyriforms		
Family	EOD[a]	Recept. Types	Family	EOD[a]	Recept. Types
Sternopygidae	W	P, T	Gymnarchidae	W	O, S
Apteronotidae	W	P, T	Mormyridae	P	K, D
Rhamphichthyidae	P	B, M			
Hypopomidae	P	B, M			
Electrophoridae	P	?			
Gymnotidae	P	B, M			

[a] EOD Electric organ discharge; *W* wave type; *P* pulse type.

invaginations of the apical surface. The gymnarchomast type I (S unit) resembles the mormyromast by its two kinds of sensory cells wich are, in contrast to those of the mormyromast, innervated by the same nerve fiber. Functionally, however, the S unit corresponds to the Knollenorgane by its spike-like activity that is correlated 1:1 with the fish's constant-frequency wave EOD, and occurs also spontaneously at similar frequency in a silenced fish (Szabo 1962a, b; Szabo and Fessard 1974).

2.1.3 Modes of Encoding Electrical Stimuli

Ampullary receptors of gymnotiforms and mormyrids are very similar both structurally and in their physiological properties. Ampullary receptor fibers fire permanently (spontaneously), sometimes at more than 100 spikes/s (around 50/s in mormyrids). An anodal step stimulus applied to the receptor opening increases the rate of firing, a cathodal stimulus decreases it (Fig. 2.8). Their response pattern therefore is opposite compared with that of the ampullary receptors of elasmobranchs (ampullae of Lorenzini), or the ampullary receptors of nonteleost bony fishes. Teleost ampullary receptors are also sensitive to mechanical stimulation, at least those of mormyrids (Szabo 1970b), as are their elasmobranch counterparts (review Murray 1974).

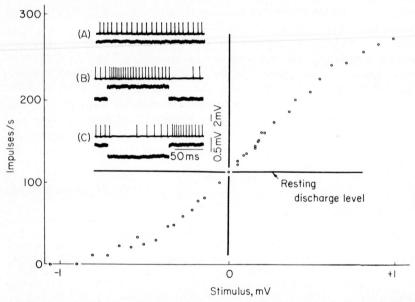

Fig. 2.8. Response of an ampullary receptor of *Gymnotus*. *A* This afferent nerve fiber discharges spontaneously at more than 100/s; its rate is modulated up or down by a *B* anodal or *C* cathodal stimulus potential. *Upper traces* impulses in the afferent nerve fiber; *lower traces* stimulating potential applied externally at the receptor opening. The stimulus/response relationship for a 100-ms stimulus is of sigmoid shape, being linear at small stimuli of either sign. Changes in sensory nerve discharge occur to changes in stimulus voltage of several µV only (Bennett 1971b)

The apical face of the cell membrane of the teleost ampullary receptor has a larger surface area than the basal membrane and a lower resistance. Therefore, the basal membrane limits the current flow through this cell. A stimulus that is anodal outside (lumenal face) depolarizes the basal face of the receptor cell which is presynaptic to a nerve fiber. The depolarization leads to an increased secretion of transmitter by the receptor cell which in turn causes the nerve fiber to fire at an increased rate. Cathodal stimuli that hyperpolarize the inner receptor cell face decrease the secretion of transmitter, thereby lowering the rate of afferent nerve impulses. These electroreceptor cells are voltage to chemical transducers and are believed to have voltage-sensitive Ca-channels that are responsible for transmitter-release or their regenerative voltage response (Bennett and Obara 1986).

The roles of the apical and basal receptor membranes have been inferred from whole-ampulla studies only (except in the transparent catfish, *Kryptopterus*; Bennett 1971b). It is believed that in the teleost ampullary receptor the physical stimulus itself, if greater than about 10 µV, causes transmitter release (linear behavior of the receptor membrane). This is in contrast to the marine elasmobranchs, where both receptor faces seem to interact, resulting in high amplification of the physical stimulus, which would explain the extraordinary sensitivity of marine elasmobranchs compared with freshwater teleosts (5 nV · cm^{-1} measured behaviorally; 10-100 times stronger threshold stimuli were determined neurophysiologically; see the review by Kalmijn 1988).

Because of their relatively slow accommodation, ampullary receptors are called "tonic". These receptors have their highest sensitivity to sinusoidal stimuli of 30-50 Hz (Poppele and Bennett; in Bennett 1971b), or 10-30 Hz (review of Zakon 1986), and thus do not respond well to DC current nor to

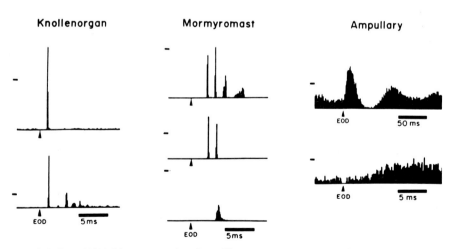

Fig. 2.9. Post-EOD histograms showing afferent nerve responses of the three types of electroreceptors to a mormyrid's own EOD ($N=100\text{-}400$). Two different Knollenorgane and three different mormyromasts are shown. The responses of an ampullary fiber are also shown at high sweep speed to facilitate the comparison with the other two receptors (Bell 1986)

high-frequency EODs. They may, however, respond to EODs if they are either low in frequency (as in certain sternopygids; Gymnotiformes) or of the pulse type (Fig. 2.9). Most pulse EODs have considerable spectral amplitudes at low frequencies. The frequency range of highest sensitivity of elasmobranch ampullary receptors is around 6-8 Hz; there is thus an additional functional difference from teleosts (reviews Kalmijn 1988; Zakon 1986).

The function of the ampullary receptors appears to be the coding of external low-frequency signals such as those generated by all aquatic animals and various nonbiological sources. A fish can "transform" a pure DC field to a low-frequency signal by its movement (see Kalmijn 1988).

The Knollenorgane respond to EODs as do mormyromasts, but Knollenorgane are 10-20 times more sensitive. They respond to an EOD by just a single spike, the latency of which remains nearly constant even if the intensity is increased (Fig. 2.9). The biological function of the Knollenorgane appears to be the detection of other fish. Their tuning to the fish's spectral EOD properties is, however, only very broad.

The mormyromasts have a wider though still narrow working range of up to about three times stimulus threshold (Kramer-Feil 1976) centered on the (strong) intensity of their own EOD. They are only weakly tuned to the broadband spectral frequency range of the EOD. They code for intensity variations induced by the presence of a conducting or non-conducting object in the fish's field if sufficiently close to the mormyromast studied. The threshold response is one spike at long latency; with increasing stimulus intensity the latency decreases and, depending on the mormyromast, more spikes may be added (up to a total of nine spikes, Kramer-Feil 1976; Fig. 2.10). The presence of objects evokes similar changes.

Bell (1990) found physiological differences between the afferent fibers arising from the two types of mormyromast sensory cells when recorded near their separate central terminations in the ELL (as shown by Bell et al. 1989; see previous Section). Chief among these were a higher threshold and a reduced number of maximal spikes from fibers arising from the outer sensory cells compared to fibers arising from the inner sensory cells. Similar findings were obtained from recordings from the peripheral electrosensory nerve, so that most likely fibers from both sensory cell types were represented in the sample.

Mormyromasts appear to be the electroreceptors enabling a fish to sense nearby objects by the distortions they impose on the fish's own EOD field (active electrolocation). However, mormyromasts mediate a specific communication function in addition to their electrolocation function: they are the receptors mediating the "echo" response (Russell et al. 1974), also called the "Preferred Latency Response" (Bauer and Kramer 1974; Kramer 1974) to another fish's EOD within a distance of up to about 25 cm, explained in Section 4.1.1.1.

The PLR's function is not yet determined, but a role in sensory gating has to be considered from Kramer-Feil's (1976) work on adaptation of mormyromast afferent responses to the second of a pair of stimulus pulses. Inhibition of the first afferent spike (and all subsequent ones, if present in that

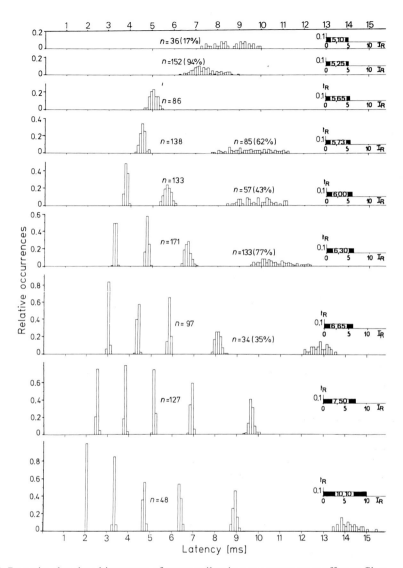

Fig. 2.10. Poststimulus time histogram of nerve spikes in a mormyromast afferent fiber (from the dorsal branch of the posterior lateral line nerve; *Gnathonemus petersii*). This fiber responds by up to six spikes; note a dramatic shortening of the latency of the first and successive spikes at increased stimulus intensity. *Ordinates* show relative occurrences for each spike; *abscissa* the poststimulus latency (*bin width* 0.1 ms). N number of responses for each spike order (in contrast to all earlier spikes, the last spike was evoked by less than 100% of the stimuli, as indicated). *Insets* stimulus parameters: t_R stimulus duration, 0.1 ms; I_R stimulus current in µA; stimulus repetition rate, 0.77/s. The stimulus was applied to the receptor opening by a local disc-shaped Ag/AgCl-electrode (surface parallel to the skin; electrode back and support insulated; disc diameter, 5.5 mm). The EOD was blocked by Flaxedil; the fish was only lightly anaesthetized by MS222 (Kramer-Feil 1976)

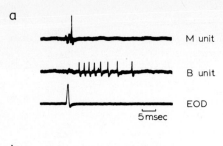

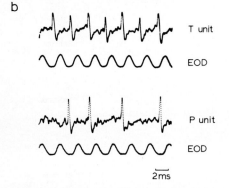

Fig. 2.11. Responses of tuberous electroreceptors in gymnotids. *a* Afferent nerve responses in a pulse species, *Hypopomus*; *b* in a wave species, *Eigenmannia* (J. Bastian in Heiligenberg 1977)

fiber) was observed when two suprathreshold, identical square wave stimuli of 0.3 ms duration were separated by up to 8 ms. For any subsequent afferent spikes, if present (see Fig. 2.10), inhibition was observed even for interstimulus intervals of up to 165 ms. Bell (1989) presents strong arguments for a single- (first-) spike latency code of intensity measurement in mormyromasts, the significance of the additional afferent spikes, if present, being unclear. The PLR of about 12 ms would assure that mormyromasts respond by at least one afferent spike to the fish's own discharge, providing it with a stable sensory input even in the presence of the discharges of another mormyrid, so long as the spatial geometry of the fish's impedance load does not change.

It is probably the Knollenorgane which enable a fish to find (locate) a congener at some considerable distance (depending on conditions, up to 1.35 m in a small mormyrid; Squire and Moller 1982) by following the current lines of its dipole field (Schluger and Hopkins 1987; see Sect. 2.3.2).

In pulse gymnotiforms, two kinds of tuberous receptors are distinguishable both physiologically and anatomically (see above). M units (pulse markers) fire a single spike in response to an EOD with little latency shift as intensity changes (Fig. 2.11). They thus signal the occurrence of an EOD but do not give information about its intensity. B units (burst duration coders) fire a variable number of spikes from 1 to 20. The latencies as well as the number of spikes depend on stimulus amplitude; the timing of the spikes relative to the EOD stimulus is less precise than that of M units. Some units are sharply tuned to the spectral frequency of peak amplitude of the EOD, others deviate considerably from it, while still others show broad tuning.

Wave gymnotiforms also have two types of tuberous electroreceptors, in general sharply tuned to their EOD frequency (compare with Figs. 4.59, 4.60).

P units show increased sensitivity also to very low frequencies and to amplitude modulation. T units (phase coders) show a strict 1:1 relationship to the fish's own EOD, which is nearly intensity-independent. P units (probability coders), on the other hand, sometimes miss one cycle and show jitter in their phase relationship to the EOD. When the EOD intensity is raised their firing rate increases (Fig. 2.11).

Both pulse and wave gymnotiforms could, in principle at least, test resistive and capacitive reactance properties of an object in their electric field (Meyer 1982) by evaluating phase shifts of local receptors (M or T units) relative to the majority, and local amplitude variation of the EOD (B or P units; Scheich et al. 1973; Scheich and Bullock 1974; reviews Zakon 1986, Bastian 1986). The mechanism of how "own" (reafferences) from "foreign" EODs (exafferences) is discriminated continues to be the subject of much discussion (Scheich 1977a,b,c; Heiligenberg and Bastian 1984; Hopkins and Westby 1986; Heiligenberg 1988; see Sects. 4.2.2.1 and 4.2.2.2).

2.1.4 Central Projections of Electroreceptive Afferents (Mormyridae and Gymnotiformes)

Recent reviews are those from Bell and Szabo (1986) and Bell (1986, 1989) on mormyrids, and Scheich and Ebbesson (1983) and Carr and Maler (1986) on gymnotiforms; see also the classic review by Szabo and Fessard (1974).

Mormyridae. For about 100 years the mormyrid brain has been a favorite neuroanatomical study object, because of its unusual size for a teleost, especially of its huge cerebellum, covering the rest of the brain ("gigantocerebellum", Nieuwenhuys; "particularly impressive" among vertebrates, Braitenberg 1977). The relationship of brain to bodyweight is nearly the same as for humans, about 1:50 (Fessard 1958). It appears that most of this "hypertrophy" is related to processing electroreceptive information.

The lateral line nerves (see above) innervate cutaneous sensory organs (like the electroreceptors, but also mechanoreceptive neuromasts, etc.). Primary electrosensory afferents terminate in the posterior part of the rhombencephalon, the electrosensory lateral line lobe (ELL). This medullary roof structure has a nucleus and a cerebellum-like cortex (see Fig. 2.12).

Primary afferents from the three types of mormyrid electroreceptors end in different regions of the ELL: for afferents from the Knollenorgane, it is the nucleus; for afferents from ampullary electroreceptors, it is the ventrolateral zone of the ELL cortex; afferents from mormyromasts terminate in both the medial and dorsolateral zones. The projections of each kind of afferent forms a somatotopic map (there are even two maps for mormyromast afferents, as indicated).

Axons of the nucleus of the ELL (the Knollenorgan pathway) project via the lemniscus lateralis to the anterior exterolateral nucleus of the torus semicircularis. The torus is a huge coordinative center of the midbrain (mesencephalon). Up to this point fiber diameters are large and synapses electrotoni-

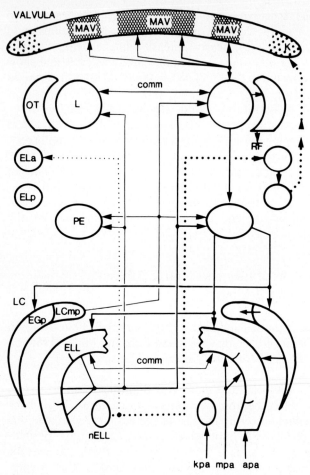

Fig. 2.12. Electrosensory connections in the brain of a mormyrid (incoming primary afferences, *lower right*). Connections of the mormyromast-ampullary system are shown with *solid lines*; of the Knollenorgan system, with *dotted lines*. The *thicker lines* indicate connections that are judged to be most important. *apa* Ampullary primary afferents; *comm* commissural connection; *K* Knollenorgan area of the valvula; *kpa* Knollenorgan primary afferents; *LC* caudal cerebellar lobe; *mpa* mormyromast primary afferents; *nELL* nucleus of the electrosensory lateral line lobe; *RF* reticular formation; *MAV* mormyromast ampullary area of valvula; *OT* optic tectum; *L* lateral toral nucleus; *ELa* anterior exterolateral toral nucleus; *ELp* posterior exterolateral toral nucleus; *PE* preeminential nucleus; *EGp* eminentia granularis posterior; *LCmp* posterior molecular zone of caudal cerebellar lobe; *ELL* electrosensory lateral line lobe (Bell and Szabo 1986)

cal, important features of a fast pathway preserving precise timing relative to a stimulus, both in electrolocation and communication. From there on fibers project exclusively to the posterior exterolateral toral nucleus, which projects, through other midbrain nuclei, to the valvula of the cerebellum.

All three cortical zones of the ELL (that is, afferents from ampullary receptors and mormyromasts) project to the lateral toral nucleus in the mesen-

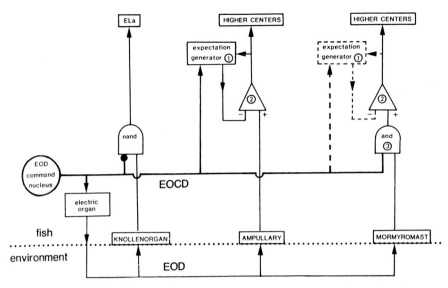

Fig. 2.13. Interactions between EOD-evoked reafferent input and the electric organ corollary discharge *(EOCD)* for all three electroreceptor systems of a mormyrid. When the *EOD command nucleus* at left evokes an EOD, an EOCD signal is relayed to the electrosensory lateral line lobe, ELL (symbolized here by components between *"higher centers"* and *horizontal EOCD-line*). In the *Knollenorgan* region, reafferent input is blocked by the EOCD (as symbolized by an inhibitory input to a *NAND gate*). In the *ampullary* region, a "central expectation" concerning the reafferent input based on past input is stored in an EOCD-driven *expectation-generator (1)*. The EOCD-driven expectation is subtracted from the actual reafferent input in the ELL *(2)*. In the *mormyromast* region of the ELL, reafferent input is facilitated by an EOCD, as symbolized by a fixed facilitatory input to an *AND gate (3)*. A subsequent adaptive process similar to that in the ampullary region has now been identified also for mormyromast input *(dashed lines)*. *NELL* Nucleus of the electrosensory lateral line lobe (Bell 1989)

cephalon, suggesting that ampullary and mormyromast afferences become integrated at a higher brain level. There are two major, somatotopically organized projections from other brain centers to the ELL, suggesting that the results of electrosensory information processing at mesencephalic and cerebellar levels influences the initial stage of information processing in the ELL. From a functional point of view the extensive, somatotopically organized coupling across the midline of electrosensory information is also interesting, suggesting that there may be a central comparison of right/left side differences.

For the processing of electrosensory information the motor command for an electric organ discharge (EOD) also has a prominent role, because corollary discharges, or efference copies, are relayed to the ELL. Corollary discharges affect the EOD-evoked, reafferent input from ampullary receptors, mormyromasts, and Knollenorgane.

As far as we know, the response of the Knollenorgane to the fish's own EOD (a reafference) conveys no useful information but interferes instead with sensing EODs from other fish. The corollary discharge of the electric organ motor command, EOCD, completely "blanks" the information from the Knollenorgan response to the fish's own EOD, by inhibiting the postsynaptic cells on which Knollenorgan afferents project (Fig. 2.13). This blanking occurs during a short period (3 ms) only, which is matched in latency and duration to the reafferent signals.

In contrast, the reafferent response from mormyromasts is enhanced by a facilitatory EOCD which also corresponds to the reafferent response in latency and duration. The function of active electrolocation is thus facilitated by an EOCD which is inhibitory in other ELL regions (Knollenorgane and ampullary receptors). There seems to be an ingenious mechanism of lateral inhibition by efferent spikes in the afferent mormyromast fibers, cancelling afferent nerve responses by "collision", perhaps aiding in contrasting an electrosensory "image" of an object from the background.

Gymnotiformes. The gymnotiform brain was studied mainly in context with active electrolocation (review Bastian 1986) and jamming avoidance responses (reviews Heiligenberg 1986, 1988). Recent reviews of the brain's functional anatomy are those from Scheich and Ebbesson (1983) and Carr and Maler (1986), but see also Szabo (1967) and Szabo and Fessard (1974).

The primary afferents innervating ampullary and tuberous electroreceptors have their cell bodies in the anterior lateral line nerve ganglion and terminate, like in the Mormyriformes, in the ipsilateral lateral line lobe (ELL) of the medulla (see Fig. 2.14).

The ELL's ventromedial segment receives input from ampullary afferents only. The other three divisions of the ELL receive input from both P and T units (*Eigenmannia*) forming three identical somatotopical maps. T and P units terminate on distinct cell types. There are four cell types projecting to the midbrain torus, the nucleus praeeminentialis (of the metencephalon), or to the contralateral ELL. Ascending efferent projections form the lateral lemniscus fiber bundle that transmits topographically ordered electrosensory information to both the torus and the nucleus praeeminentialis. The time coding T units project only to the torus (see below).

Similarly as in the mormyridae, there is an electrosensory information loop within the metencephalon. Information runs from the nucleus praeeminentialis to the cerebellar caudal lobe (which also receives mechanoreceptive and proprioceptive input). From there, information is returned to the ELL (although other connections are known as well, for example, to the tegmentum and the torus).

The torus seems to be the main site of electrosensory information processing. It is a huge, "superlaminated" (Scheich and Ebbesson 1983) midbrain structure unlike anything found in other teleosts (although the torus of mormyrids is also large it is not laminated). The torus is the equivalent of the inferior colliculus in mammals; it has 12 laminae and about 48 cell types (15 laminae according to Scheich and Ebbesson 1983).

The torus receives lemniscal electrosensory, mechanoreceptive, and auditory input; its huge size is in association with the electrosensory input which is confined to the dorsal torus, while auditory and mechanoreceptive

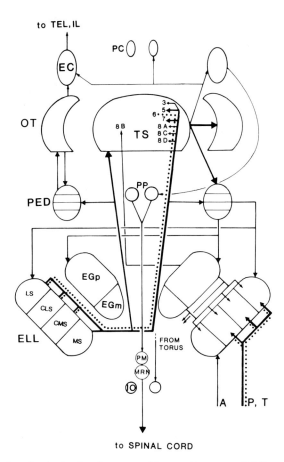

Fig. 2.14. Electrosensory connections in the brain of a gymnotiform with a wave EOD, *Eigenmannia*. Primary afferents terminate in the *ELL*, electrosensory lateral line lobe, and are of the *A* ampullary, *P*- and *T*-type *(lower right)*. *Heavy lines* lemniscal system (connections to higher brain centers). In the lemniscal system, *dotted lines* denote the *T*-type efferents of the ELL; *intact lines* the ampullary and *P*-type efferents. The laminar organization of the *TS* torus semicircularis is shown by numbers. *TEL* telencephalon; *IL* inferior lobe of diencephalon; *PC* posterior commissure; *OT* optic tectum; *PED* dorsal preeminential nucleus; *PP* prepacemaker nucleus; *EGm, EGp* eminentia granularis medialis and posterior. The four segments of the ELL: *MS* medial, *CMS* centromedial, *CLS* centrolateral, *LS* lateral. *PM* pacemaker; *MRN* medullary relay nucleus; *IO* inferior olivary nucleus (Carr and Maler 1986)

systems project to the ventral torus. Certain neurons of the dorsal torus respond to various combinations of amplitude and phase (timing) of electrical stimuli; for example, those which occur during the presentation of objects (electrolocation context) or the presence of real or simulated conspecifics (electrocommunication context).

In the torus the somatotopically organized electrosensory information is preserved in many of its laminae. The torus is the site of integration of the electrosensory system with the visual system, with motor control, and with electrosensory feedback to lower electrosensory nuclei.

Examples of ascending electrosensory projections from the torus are the ones to the optic tectum and the nucleus electrosensorius (located at the diencephalic pretectal border). The nucleus electrosensorius receives a bilateral projection from the torus and a unilateral one from the tectum. Among several projections originating from the nucleus electrosensorius the descending one to the prepacemaker nucleus may be of greatest significance for the control of the jamming avoidance response (Sect. 4.2.2.2). The paired prepacemaker nucleus may be the only afferent input to the medullary pacemaker nucleus, a single midline nucleus composed of electrically coupled cells that control the discharge frequency of the electric organ (see Chap. 2.2.2).

Some electrosensory brain centers and areas of sensorimotor integration contain monoamines, a class of transmitter substances including noradrenaline (NA), dopamine (DA) and serotonine (5-HT), which are known to be involved in the reproductive function and motor control (DA), the regulation of arousal and mood (NA, 5-HT), and sensory perception (5-HT). In *Eigenmannia*, all the main layers receiving electrosensory input, especially the large T-cells of the torus' layer 6 that are part of the fast-conducting, phase-sensitive system, are monoamine-perikarya. Both types of cells of the medullary pacemaker nucleus, the pacemaker and the relay cells, contain monoamines (Bonn and Kramer 1987). The only functional evidence to date concerning a possible role of monoamines in the electromotor system is the observation that chlorpromazine, a dopamine antagonist, reversibly reduces the EOD frequency in *Apteronotus* (Kramer 1984).

2.2 Electric Organs

Electric organs are specialized to generate an electric field around a fish. Electric organs are only found in certain cartilaginous fishes (rays and skates, or Batoidea), and certain teleosts (Fig. 2.15). Skates and rays are either weak or strong electric, as is also true in electric teleosts (with one species, the electric eel, being both).

Strong electric fish, such as the electric ray, *Torpedo*, of the Mediterranean, the electric catfish from the Nile, and the electric eel of tropical South America, were certainly known to primitive man and have their place in ancient art (the catfish, but also certain mormyrids, is depicted on Egyptian murals), ancient Greek and Roman medicine (*Torpedo*), belletristic literature (Balzac reporting upon the voyages of A. v. Humboldt and his encountering the eel in South America), and comics magazines of our time (adventures of the French/Belgian Tintin). In the second half of the eighteenth century, strong electric fish played an important part in the development of the physi-

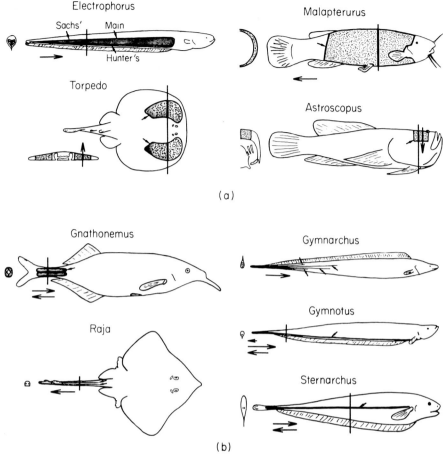

Fig. 2.15. Representative electric fish. *a* Strongly electric; *b* weakly electric. All are shown from the side except *Torpedo* and *Raja*, which are shown from the top. Electric organs indicated by *stippling* and *arrows*. Cross-sectional plane as indicated by *line*. The *large arrows* indicate the direction and sequence of current flows through the organs; the length of these arrows is proportional to the amplitude of the successive phases (if there is more than one). *Raja* and *Torpedo* are cartilaginous fishes; all other fishes are teleosts. *Astroscopus*, the stargazer, is a perciform; *Malapterurus*, the electric catfish (Siluriformes); *Gnathonemus* (a mormyrid) and *Gymnarchus* are mormyriforms; *Electrophorus*, the electric eel, *Gymnotus* and *Sternarchus* (an apteronotid) are all gymnotiforms (Bennett 1971a)

cal science of electricity, and in electrophysiology (reviews Kellaway 1946, Wu 1984, Zimmermann 1985).

In contrast to weak electric fish, strong electric fish discharge only intermittently. Probably all strong electric fish feed mainly or exclusively on other fish, that is, extremely sensitive and agile prey that are exceedingly difficult to catch. An organ discharge volley at the moment of attack greatly increases the success rate, apparently by numbing or panicking the victims (*offensive role* of discharges). Another established function of the strong organ is *defense*. All strong electric fish that have been studied emit discharge volleys when prodded with a stick. Strong organs are not known, however, to play an important role in *intraspecific communication*, although so little is known about their behavior, especially during reproduction, that this cannot be excluded. (There is some evidence for a social role of strong discharges in feeding eels; Bullock 1969). Active electric catfish discharge not only during attack and defense, but apparently also to stir up and chase prey fish (*Scheuchentladungen; beating function* of strong discharges), although the catfish rarely discharges at all in an intraspecific context (see Sect. 3.3.3).

Strong discharges could, in principle, also serve in *active electrolocation*, although there is, to date, no clear evidence for such a role. All strong electric fish, except the "highest", the stargazer (Perciformes), possess ampullary electroreceptors that may be expected to respond to the fishes' discharges. Unlike most weak electric fishes, strong electric fishes emit pulsed DC current (which perhaps leads to more effective shocking of the prey or aggressor). Only the eel, like its fellow-gymnotiforms, possesses in addition tuberous electroreceptors that are sensitive to the higher frequency components of discharges. (There is recent field and experimental evidence suggesting that the eel may prey largely on other electric fish; Westby 1988).

Weak organ discharges, on the other hand, are only a recent discovery (Lissmann 1951, 1958; Coates et al. 1954; Grundfest 1957), although weak electric organs of all major groups were well-studied anatomically by the turn of the century. Sometimes weak organs were referred to as "pseudoelectric" (because of their histological similarity to strong organs). A human observer handling weak electric fish would normally never notice the feeble currents they generate, and electronic equipment is needed to detect the discharges (except in certain species, like *Mormyrus rume* of sufficient size; see Chap. 3.1).

All weak electric species discharge continuously during their whole life; short pauses in response to disturbances may occur but are relatively rare, except in social context. The function of weak organs are *active electrolocation* and *communication* (both intra- and interspecific); weak organs are not known to serve an offensive or defensive role like strong organs.

Darwin discussed the problem of how a strong electric organ, like that of the eel or of *Torpedo*, could have evolved by small intermediate steps when the initial stages could have had no adaptive advantage (in "The Origin of Species", chapter "Difficulties of the Theory"; see Bennett 1971a). Evidence for an electrosensory function of the transition stage that was selected for came only 100 years later (Lissmann 1958). Although certainly a key discovery and part of the solution for a longstanding evolutionary riddle, this explanation might also suggest that weak electric fish will *all* become strong electric fish, given enough time.

Who would want a weak rather than a strong organ? Where is the adaptive advantage of a weak organ? Communication and electrolocation should be possible both with a weak and a strong organ; if not rather better with a strong one.

It appears that in most cases, a weak organ is not simply an example of incipient evolution, but represents a different "evolutionary strategy", or adaptive peak (for an introduction into this terminology, see, for example, Krebs and Davies 1987). By "giving away" the option of being powerful and terrifying (with easy access to a fish meal), weak electric fishes gain improved capacities to actively and constantly probe their nocturnal environment, and to exchange information with conspecifics via a private channel. These gains may compensate fully for the losses; there are quite a few weak electric fishes that are heavily or exclusively piscivorous; for example, the (in general well-fed) huge *Gymnarchus*, which could easily accommodate the tissue for a strong organ. An active location system is only known for certain birds and mammals (echolocation or SONAR), and has evolved several times independently in these "highest" land-living vertebrates. Some of these systems are incredibly sophisticated and exceedingly powerful location and discrimiation devices (for example, in certain bats; Griffin 1958; Neuweiler 1984; von der Emde and Menne 1989).

There are two reasons why an organ designed for electrolocation and communication, that is, continuous operation, must be weak: (1) The energy cost of a continuous strong field would be prohibitively high. Continuously discharging fish all have weak organs, and have exclusively evolved in tropical freshwater biotopes with their low ion content (that is, high water resistivity) that keeps energy requirements low. Some marine skates, for example, are weak electric; they discharge, however, only intermittently (for example, Bratton and Ayers 1987). (2) Too strong an organ would be counterproductive because it would signal the presence of the sender also to non-electroreceptive prey and predators (by stimulating their nerves directly and making their muscles twitch), giving them a headstart. Not only eavesdropping of the weak electric fields generated by all aquatic organisms, but also private communication would not be possible with a continuously operating, strong organ.

From a biogeographical viewpoint one may wonder why there are no freshwater electric fish in tropical South East Asia, a region famous for its especially rich vertebrate radiations; in general, richer than those of Africa or South America (Fittkau 1985) where we do find distinctive electric fish communities.

2.2.1 Structure and Function of Electric Organs

The study of electric organs has contributed significantly to membrane physiology. Histology and ultrastructure of electric organs, electrophysiology and ionic mechanisms, biochemical aspects and transmitter mechanisms have been studied in depth. For the purposes of this book, Bennett's (1971a) comparative study is the most important reference (and should be consulted for more details and references to the original work). Bennett's study is a definite treatise of most aspects of the subject; a recent review is Bass (1986).

Electric organs are derived from muscle tissue (there is one exception, see below), although different muscle groups are involved in different taxa (see Fig. 2.15). These muscle cells are unusual in that they do not twitch when neurally excited by transmitter substance (acetylcholine); various ultrastructural anomalies have been found in different groups, which may explain why in electric organs the electromechanical coupling of normal muscle cells does not work (for the mechanical part).

Often these muscle cells are short cylinders (resembling coins) that are stacked one upon another, forming one of several, parallel columns enclosed in a tight jacket of connective tissue. There is also connective tissue inside the columns, as well as blood vessels and nerves. Sometimes these cells are long and spindle-like (for example, in *Eigenmannia*); therefore, the name "electrocyte" (Bennett 1971a) seems more appropriate than some earlier names [electroplate(s), electroplax(es), electroplaque(s)].

There is an inherent problem for all electric fish: the voltage generated by one active cell tends to hyperpolarize the innervated face of the next cell in series with it and thus prevent its firing. Therefore, each electrocyte must be innervated separately to receive the central command synchronously, by a spinal motoneuron that may form part of a nucleus (mormyrids). The presynaptic fiber may arborize and contact the cell multiply, or there may be one or several stalks of the electrocyte that are contacted by the endings of the nerve fiber. An axon of a medullary relay cell contacts each electromotoneuron in mormyriforms and gymnotiforms. The cells of a medullary command nucleus "decide" when to fire the electric organ; all electrocytes are excited synchronously. Because the electrocytes of one column are in series, their potentials add. In general the columns are oriented rostro-caudally; so is the potential difference (and the direction of internal current-flow) when the organ is fired.

In the bottom-dwelling stargazer and the electric ray the columns are oriented vertically (dorso-ventrally), in accordance with their upwards directed attacks on prey fish (Belbenoit and Bauer 1972 for *Torpedo*; Pickens and McFarland 1964 for the stargazer).

The marine species with strong organs (with the stargazer as the only teleost) have flattened organs with many columns in parallel (500-1000 columns in parallel, each with about 1000 cells in series in *Torpedo*; 150-200 cells in series in the stargazer, a "weak" strong electric fish). Their organs generate a low voltage, high current output as is adequate for their conductive medium (*Torpedo*: the pulse amplitude is 50 V in air, with a power output exceeding 1 kW at the peak of the pulse). In contrast, electric organs of freshwater species (teleosts) are often long, generating a high voltage, low current field as indeed they must in a medium of high resistivity. The eel has about 6000 electrocytes in series, and dorsoventrally about 35 (bilaterally) in parallel.

In contrast to all other electric fish, the apteronotids (of the gymnotiforms) have neurogenic electric organs; their presynaptic nerve fibers have lost their contact with muscle cells and form the organ. (Larval apteronotids have a temporary organ of myogenic origin; Kirschbaum 1983.) The apteronotids show species differences in discharge waveform that are related to organ morphology and physiology; apteronotids are outstanding for their very high discharge frequency, which reaches 1800/s in certain species. No ordinary nerve

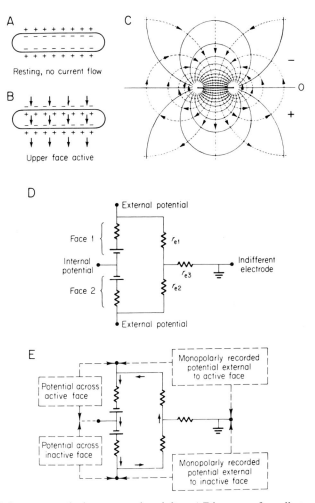

Fig. 2.16. Equivalents of electrocytes during rest and activity. *A* Diagram of a cell at rest; equal potentials are opposed. *B* When the upper face generates an overshooting action potential, two potentials act in the same direction and current flows as indicated by the arrows. *C* Equipotentials (*solid lines*) and lines of current flow (*dotted lines*) around a thin electrocyte indicated by the *heavy horizontal line*. The isopotentials are separated by equal increments and meet at the two edges of the cell. *D* Electrical equivalent of a resting cell. The resistance of the external current path can be represented by re_1, re_2, and re_3. *E* Equivalent of an active cell. The *small arrows* indicate the direction of current flow (Bennett 1971a)

or muscle tissue comes close to even half that rate (at least not in sustained activity). These exceedingly high frequencies may only be possible because of a command pathway with entirely electrotonical synapses, and the specialized anatomy and physiology of the organ (see also Waxman et al. 1972). But certain sternopygids, for example of the genus *Rhabdolichops*, may also discharge at greater than 800/s, although the motor synapses of the organ are chemical, and the effector organ is specialized muscle tissue.

43

All aquatic organisms create electric fields around them, sometimes of considerable strength (see, for example, Kalmijn 1988), but electric fish have anatomical and physiological adaptations that increase the strength of the fields, and that control for a precise onset and synchronization of the excitation of the electrocytes, and high stability of waveform and often frequency. Otherwise, the system operates on the same general principles as ordinary nerve or muscle (explained in Fig. 2.16).

One of the mechanisms for precise synchronization concerns the compensation for conduction time differences from the medullary pacemaker to electrocytes located in near or far parts of the organ (the electric eel may reach 2.5 m in length, requiring synchronization among electrocytes about 2m apart). Fibers running to nearer parts of the organ take a more devious path, or have a lower conduction velocity than fibers running to the far end of the organ (see Bennett 1971a; Meszler et al. 1974).

The ionic mechanisms of electrocyte membranes differ widely among species; these differences are the main source of the wide variation of organ discharge waveforms and frequencies among species. An organ discharge may be either of the pulse ("buzzer") or of the wave ("hummer") type (see Fig. 3.1); they are called hummers or buzzers because of the sound of their audio-amplified discharges. Wave discharges are all weak.

The mechanism of the eel's discharge was the first to be elucidated (Keynes and Martins-Ferreira 1953; Altamirano et al. 1953). The eel emits weak and strong pulses (about 10 and 500 V or more in a strong fish). The weak pulses are emitted continuously, at a few per minute when the animal is resting, and at about 30/s when it is actively swimming. The weak pulses are believed to subserve active electrolocation (perhaps also communication), the strong ones are emitted when the fish attacks its prey, or in defense.

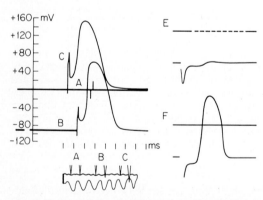

Fig. 2.17. Responses of electrocytes of the electric eel. *A* Both electrodes external to the innervated face; no response to a brief stimulus (there is a diphasic artifact) is seen. *B* One electrode is advanced into the cell. The inside negative resting potential of about 90 mV and an overshooting action potential of about 140 mV in amplitude are recorded. *C* When the exploring electrode is advanced to outside the uninnervated face, the resting potential disappears, but the spike is essentially unchanged (Bennett 1971a, after Keynes and Martins-Ferreira 1953)

The weak pulses are emitted by the Sachs' organ, the strong ones by the Main organ, while the Hunter's organ seems to be functionally divided and to contribute to both. The Sachs' organ also contributes to the strong pulses in a minor way (Albe-Fessard and Chagas 1954). The discharge waveforms are the same for weak and strong pulses (monophasic, head-positive pulses of about 2 ms).

The electrocytes are innervated on their posterior faces by spinal nerves that contact the cell primarily on short stalks. The anterior, uninnervated faces have a surface considerably increased by a large number of papilli.

The innervated face responds to depolarization by an overshooting spike of unusual amplitude (150 mV; Fig. 2.17). The uninnervated face of very low resistance does not become excited (0.2 $\Omega \cdot cm^2$ as compared with 19 $\Omega \cdot cm^2$ for the innervated face, and about 3000 $\Omega \cdot cm^2$ for frog twitch muscle; review Bennett 1971a). The two faces of the electrocytes are thus fairly matched in their impedances, still more so when the innervated face becomes excited (and its resistance declines). This kind of impedance matching also comprises the extracellular space and the external environment, and is seen in many electric fish.

At rest, the cell's membrane potential is largely determined by intra- and extracellular K^+ concentrations. The inward current is Na^+-dependent, as shown by the effectiveness of the specific blocking agent tetrodotoxin. Unlike the squid axon and many other tissues, the eel's electrocytes lack a delayed K^+ outward current, which would quickly restore the resting potential. Instead, the time constant of the membrane is sufficiently short so that the cell can be fired at a rate of almost 500/s (for a brief period). As Bennett (1971a) points out, it makes sense for an organ designed for maximal power output that the circuit for all the Na^+ inward current of a cell should be completed by the external environment, and not by local opposing currents in the innervated cell membrane.

In contrast to freshwater electric fish, marine strong electric fish, including the stargazer, generate exceptionally large PSPs (of up to 90 mV amplitude) instead of spikes. Their membranes can only be excited chemically, not by depolarization. The advantage of a PSP-generating over a spike-generating membrane in the marine environment is unknown.

Weak electric freshwater fishes, such as the high-frequency apteronotids, tend to have no or little DC associated with their discharges, which allows them to have a more effectively dual electrosensory sytem: one for low-frequency voltages of primarily external origin, and another for monitoring the higher frequency organ discharges. The wave fish *Gymnarchus* (perhaps also *Eigenmannia*) achieves an organ discharge free from a DC component by modification of one electrocyte face (the uninnervated one) to pass current only capacitatively. It has a large capacity and is of high resistance and inexcitable (Bennett 1971a); for anatomical data see Srivastava and Szabo (1972, 1973).

Pulse fish, such as a biphasic *Hypopomus* species, *Gymnotus*, or the mormyrids have the opposed faces of their electrocytes act in sequence to achieve a similar effect. The uninnervated face is electrically excited to

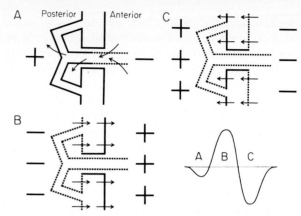

Fig. 2.18. Current flows generating triphasic pulses in electrocytes with penetrating stalks in mormyrids. Diagrams show a region near a single penetration during different stages of activity. Active membrane is indicated by *dotted outlines*. *Arrows* show direction of current flow. Resulting potential is shown on the *lower right*. *A* Head negativity is produced when the stalk activity, initiated near the site of innervation, is passing through the penetration. *B* Head positivity results when the impulse in the stalk excites the caudal face. *C* Head negativity is again produced when the rostral face becomes active (Bennett 1971a)

generate a spike that is slightly delayed compared to the spike of the innervated face. The net result is a biphasic potential, because the currents flow in opposite directions (with some cancellation in the shorter discharges).

In contrast to mormyrids, many gymnotiforms (excepting sternopygids) have more than one organ, which are either anatomically distinct (as in the eel, or in certain hypopomids and apteronotids with rostral accessory organs), or functionally heterogeneous (as in *Gymnotus* where the dorsal portion of the organ is fired ½ ms earlier than the rest; in addition to inverted polarity because of its reversed innervation). This complexity is reflected in additional phases or inflexions to the basically biphasic discharge that are species-characteristic or probably species-specific in many cases (see Chap. 3).

Mormyrids have mechanisms of their own that complicate their basically biphasic organ discharge. These mechanisms include more or less elaborate stalks of the innervated face of the electrocytes (see also Szabo 1958, 1961). The simplest (probably primitive) stage appears to be that of *Mormyrus rume* with multiple innervation on fine and numerous stalks. In species with shorter discharges the number of innervation sites is reduced to the final limit of one, suggesting that more precise synchronization can be achieved with fewer innervation sites. Synchronization is particularly important in these bi- or triphasic discharges, because slight out-of-phase firing leads to cancellation.

The stalks may, in certain species, penetrate the electrocyte and find their nerve on the "wrong", usually the anterior side of the cell, and this shows up in the overall organ discharge (Fig. 2.18). Fish possessing organs with penetrating stalks all produce triphasic discharges, the biphasic "main" discharge being preceded by a head-negative potential. There are species in which the stalks penetrate the cell twice so that they contact their nerve on the "correct", usually the posterior side of the cell (such as in species that have nonpenetrating stalks). There are even species that have a combination of penetrating with non-penetrating stalks (review Bass 1986).

In some species such as *Pollimyrus isidori* the time constants of both faces of an electrocyte seem to show a high intraspecific variability. The relatively long-lasting head-positive potential by the posterior face is "cut" into two parts by an overriding, very brief and strong spike of opposite polarity generated by the anterior face. A split-microsecond precision of the timing of both faces' activity is critical for the overall waveform and its intraspecific variability (modelled numerically by Westby 1984; experimental evidence, Bratton and Kramer 1988).

Distinct larval organs that precede the adult organ and regress on its becoming functional have been described not only in an apteronotid (see above) but also in *Eigenmannia* (another wave-gymnotiform, Sternopygidae) with a myogenic adult organ, but are not known, to date, in pulse gymnotiforms. However, mormyrids do have larval organs that emit pulse discharges, although of different waveform (head-positive, monophasic) and longer duration compared to those of the adult organ (Kirschbaum 1977, 1981; Kirschbaum and Westby 1975; Westby and Kirschbaum 1977, 1978; Denizot et al. 1978; see also Heymer and Harder 1975; Szabo 1960).

The discovery of larval organs has fostered hypotheses of the evolution of organs; no clear picture has yet emerged. The behavioral role of larval discharges also remains to be elucidated.

Bass (1986) reviews evidence for the effect of androgen hormones that, when administered experimentally, increase the duration of the discharge in some mormyrids. He also reviews some of the sex differences in gymnotiform pulse (discharge duration) and wave fish (discharge frequency), and in mormyrids (see also Chap. 3). As explained in Section 3.1.1.1, the reports about mormyrids need more support, and taxonomic problems (sibling species complexes) have to be resolved. In *Pollimyrus isidori's* discharge waveform, the best-documented case, a statistically significant difference between male and female populations is present, which, however, shows wide overlap (Fig. 3.4; Table 3.1). In *Eigenmannia* with its AC wave discharge, adult males have narrower pulses superimposed on a head-negative base-line than females (which have a more closely sinusoidal discharge; Fig. 3.15); therefore, adult male discharges have a higher overtone content (Fig. 3.16). A similar dimorphism has been suggested for *Sternopygus*.

The waveform of mormyrid pulse discharges, but not of *Eigenmannia*'s wave discharges, is sensitive to being loaded by water of different resistivity within the natural range (see Sect. 3.1).

2.2.2 Neural Control of Electric Organs

Early work is summarized in Bennett's (1971a) study. More recent reviews are those by Dye and Meyer (1986) for weakly electric teleosts, and Bell and Szabo (1986) for mormyrids. Carr and Maler (1986) touch briefly on a wave gymnotiform, *Eigenmannia*, in their review on central nervous electroreception.

Command or "pontifical" neurons have long been unpopular in neurobiology, a neural "democracy" being more timely. Electric fish, however, give beautiful examples for command neurons of their organs, or small "committees" of such (Bennett 1971a). Command neurons, where they have been found, have often been called pacemaker neurons because they proved to be spontaneously active (even in a brainslice chamber; review Dye and Meyer 1986), comparable to the pacemaker cells of the heart. All species, including the weakly electric fish, can time and modulate their discharges or discharge frequencies; therefore, excitatory and inhibitory inputs on the pacemaker must exist. Pacemaker neurons are often coupled by electrotonical synapses (sometimes called "synchronizing" synapses; Bennett 1971a).

In mormyriforms and gymnotiforms, close synchronization of the command signal in cells of one level extends down to relay cells, and from there to the spinal electromotoneurons. Coupling is achieved by contact between somata or dendrites, or by presynaptic fibers.

The electric catfish's simple command system consists of only two giant electromotoneurons (greater than 100 µm in diameter) in the first spinal segment, one on either side. Both cells are closely coupled electrotonically by presynaptic fibers, and behave as a unit functionally. One command pulse fires the organ once. Each giant cell innervates the millions of electrocytes on its side of the body (Fig. 2.19).

In gymnotiforms the command system has four levels, from peripheral to central: electromotoneurons, medullary relay cells, medullary pacemaker cells (all shown in Fig. 2.19), and mesencephalic prepacemaker cells (Fig. 2.14). Pacemaker and relay cells either form two separate, but closely adjacent, midline nuclei (for example, in the pulse fish *Hypopomus*), or are intermingled in one nucleus, as in the wave fish *Eigenmannia*.

Depending on the species, there are some 30 to 200 pacemaker cells activating about 50 large relay cells in gymnotiforms. These project to some hundreds to thousands of spinal electromotoneurons. In all gymnotiforms except apteronotids, electromotoneurons innervate a number of electrocytes. In apteronotids the electromotoneurons themselves generate the discharge. The connection of the command system to electroreceptive afferences is by the prepacemaker nucleus, which has been shown to modulate the pacemaker firing frequency by stimulation experiments (review Dye and Meyer 1986).

At each level of the gymnotiform command system a single spike occurs for each organ discharge. However, in wave gymnotiforms the electromotor neurons have been observed to continue to fire at a similar frequency after abolishing their input from relay cells by spinal section (review Dye and Meyer 1986). Ringing seems to be an intrinsic property of all parts of the command system in these fish (and has even been observed in electroreceptors), and may somehow be necessary for generating the most stable biological rhythm, the wave discharge.

In mormyrids the control system is more complex and not completely resolved (reviews Bennett 1971a; Bell and Szabo 1986). The elec-

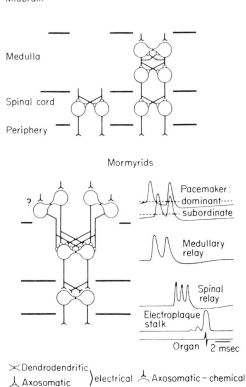

Fig. 2.19. Neural circuitry controlling electric organs in teleosts. The modes of transmission are diagrammed as shown in the key below. Where axosomatic synapses are indicated, axodendritic synapses are also found. The mode of transmission to the command nucleus is known only in the electric catfish *Malapterurus*, although it is indicated as chemically mediated in the others. The pacemaker in mormyrids (*question mark*) is now known to be a small medullary midline nucleus, not a paired structure; thus, there is no dominant and subordinate pacemaker; see Fig. 2.20. In *Malapterurus* and the gymnotiforms a single command volley at each level precedes each organ discharge. In mormyrids the activity is more complex and is diagrammed for each level. Relay cells fire in doublets that evoke a triplet of spikes in spinal electromotoneurons. The electroplaque (or electrocyte) stalk responds by three PSPs (with the second and especially third being greatly facilitated); only the third reaches firing threshold of the organ (Bennett 1971a)

tromotoneurons that innervate the electrocytes of the organ form a nucleus in the caudal spinal cord. They are driven by the cells of a medullary relay nucleus, a single midline structure, by chemical synapses. Among each other, electromotoneurons and relay neurons are coupled together electrically. The medullary relay cells fire in "doublets" that evoke a triplet of spikes in the electromotoneurons (Fig. 2.19). The three spikes are propagated out to the electrocyte stalk, where the first spike causes a small PSP, the second spike a

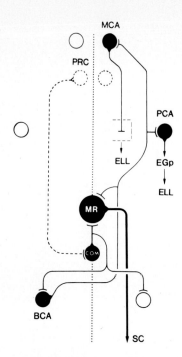

Fig. 2.20. Connections among the motor command nuclei of the organ discharge in mormyrids. The *dotted line* down the middle of the figure indicates the midline; rostral is up. The medullary command nucleus (*COM*) is the initiation site of a motor command that ultimately fires the organ. These cells receive diffuse local and also two specific inputs (both of mesencephalic origin; one shown, *PRC* precommand nucleus). Medullary relay neurons (*MR*) receive their major input from the command nucleus neurons, and send large axons out to the spinal electromotoneurons that fire the organ (*SC* spinal cord). Electroreceptive brain areas are "informed" regarding a motor command by corollary discharges that arise from collaterals of the command cells' axons (*COM*), to the bilateral bulbar command-associated nucleus (*BCA*). This nucleus' cells send information to the electrosensory lateral line lobe (*ELL*), (1) via the mesencephalic command-associated nucleus (*MCA*), and (2) via the paratrigeminal command-associated nucleus (*PCA*) and the eminentia granularis posterior (*EGp*) (Bell and Szabo 1986)

greatly facilitated PSP, and the third PSP reaches threshold. Thus each volley of three spikes (about 1 ms apart) evokes only one organ discharge. The PSP triplet can be recorded externally (Westby and Kirschbaum 1978; Bratton and Kramer 1988).

A pacemaker nucleus has only been identified recently in mormyrids (Bell et al. 1983; Grant et al. 1986; see Libouban et al. 1981 for *Gymnarchus*). This command nucleus is a midline nucleus of 16-20 relatively small neurons located just ventral to the medullary relay nucleus. The cells are coupled together by gap junctions (Elekes and Szabo 1985). In contrast to the cells of the medullary relay nucleus, their dendrites extend far beyond the confines of the nucleus, into the surrounding reticular formation and longitudinally running fiber tracts, where they are presumably contacted by the most diverse sources (Fig. 2.20). These afferent inputs probably are responsible for the sensitivity of a mormyrid's discharge rhythm to virtually all kinds of excitation. Two specific projections to the command nucleus from the mesencephalon are also known (one of them indicated in Fig. 2.20).

Axons from the command nucleus contact the medullary relay neurons by club endings (mixed electrical and chemical synapses; Elekes et al. 1985). Collaterals of these axons project caudally to a bilateral bulbar command-associated nucleus (BCA in Fig. 2.20). Probably all command-associated corollary discharges seen in the electrosensory lateral line lobe (ELL) arise here (corollary discharges "inform" afferent brain areas that a reafference, which is sensory input caused by one's own action, is expected; see Fig. 2.13). The medullary relay neurons are doubly innervated: (1) directly, by axons of the command neurons; (2) indirectly (and chemically), by collaterals of axons from the BCA. The function of this double innervation is not clear; perhaps the input from the BCA assures the second spike of the doublet that is the relay cells' command signal to the electromotoneurons (Bell and Szabo 1986), although firing in doublets appears to be an intrinsic property of the relay cell membrane (Bennett 1971a).

2.3 The Electric System in the Aquatic Environment

The adaptive advantage of the electric system, as found in weakly electric fishes, lies in the gain of specific information. In active electrolocation, information about objects nearby is obtained; that is, position, size, conductive

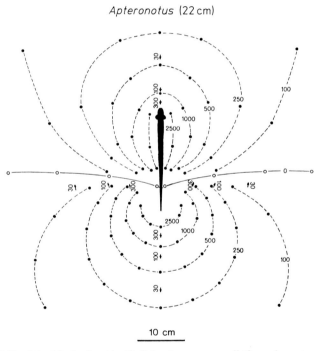

Fig. 2.21. The electric field of the black ghost knife fish, *Apteronotus albifrons*, in water of 263 µS·cm^{-1} conductivity (or 3.8 kΩ · cm resistivity), suspended just below the surface ("half space" measurement). The fish is viewed from above in the center of a circular 300 cm diameter × 90 cm deep test tank; the field was measured relative to a reference electrode 150 cm away at the zero equipotential of the fish's field, in the horizontal plane at the middorsal level. The field potential values in µV$_{peak-to-peak}$ are written horizontally next to the appropriate equipotential line; identical values for the caudal half are of opposite polarity. Field gradient vectors in µV · cm^{-1} are shown along the axis of the fish and at right angles to the fish. Field gradient magnitudes are written vertically, next to their corresponding vectors. Note the marked rostro-caudal asymmetry of the field due to the distortive effect of the body. The organ discharge is head-positive and of the high and constant-frequency, AC-wave type of about 800-1000 Hz; see Fig. 3.22 (Knudsen 1975)

properties, and movement. An "object" can also be an organism. In electrocommunication, the subject of this book, a fish may signal its species identity, age, sex or location to a conspecific, or it may broadcast its state of excitement, tendency to attack, readiness to mate, etc. In schooling species, the organ discharge aids in group cohesion (Moller 1976), just as visual color marks do in some fish (for example, many of the small Amazonian characins like the neon tetras or the cardinal tetras; Roberts 1973).

Electric organs are specialized for generating an electric field in the external environment. In the tropical freshwaters weakly electric fish live in, this is generally water of low ion content and high resistivity (usually above 10 kΩ · cm, or a conductivity below 100 µS · cm^{-1}). Thus, the electrical load into which a fish discharges its organ is small (Bell et al. 1976; see also Bratton and Kramer 1988). This is a condition that maximizes the range at which the signal is detectable while the energy expenditure is minimal (see Sect. 2.3.2).

At some distance the field generated by an electric fish corresponds to a dipole field; however, the near field deviates considerably (Fig. 2.21). Relative to a distant electrode, amplitudes, of opposite polarities, are greatest near the tail and near the head (opercula). While the tail end behaves as one would expect from a physical dipole, the front part of a fish may be symbolized by a longer electrode that expands its half of the dipole field. In the rostral part of a fish, isopotential lines are fairly parallel to the fish's skin, so that current flows perpendicular to the skin. The decay of field intensity with distance is slower in the front compared with the tail half of the field, especially in larger fish; the density of receptors is generally higher in the front part (see Sect. 2.1.1).

2.3.1 Active Electrolocation

Mormyriforms and gymnotiforms are members of an exclusive club of animals that emit their own energy to test the environment and to locate objects, making them independent from the fortuitous and often unpredictable signals the environment provides (Lissmann and Machin 1958; reviews Lissmann 1963; Scheich and Bullock 1974; Heiligenberg 1977; Bullock 1982; Scheich 1982; Bastian 1986). Contrary to most animals that exclusively depend on environmental signals, weakly electric fish control the energy form, the intensity, and the frequency of their signal; organs and receptors are coadapted so that the system functions near its optimum.

During the pioneering years of the study of weakly electric fishes (late 1950s and 1960s), the electrolocation system received almost exclusive attention compared with the electrocommunication function that was only prophesied, but not studied until about 1970. Studies of the peripheral and central nervous structure and function dominated the scene. Electrolocation *behavior* has, however, not been studied in very great detail even at the time of this writing. No very clear picture about its biological relevance is yet available. We understand the effects of good and bad conductors (that is, metal and plastic) on a fish's field and peripheral and central nervous afferent input;

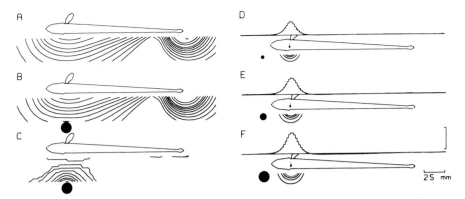

Fig. 2.22. The electric dipole field generated by the organ of the brown ghost knife fish, *Apteronotus leptorhynchus*, in water of 100 µS · cm^{-1} conductivity (or 10 kΩ · cm resistivity). Only the left half of isopotential contours is shown. Fields measured with *A* no distorting objects in the water (*similar to Fig. 2.21*) and *B* with a 12-mm diameter, 5-cm-long metal cylinder. Note compression of field lines between metal object and fish, corresponding to an increase of the field gradient normal to the fish's skin. (An insulator would have opposite effects). Contour values beginning closest to the fish are 17, 15, and 13-6 mV$_{p-p}$ in 1-mV steps. *C* The difference field (B-A). *D-F Upper traces* change in transepidermal voltage, ΔV, measured at the position of the *arrow* as a function of rostro-caudal object position for 4, 8, and 12 mm metal cylinders positioned 5 mm lateral to the fish. The *vertical tic mark* in *F* corresponds to 2.5 mV and holds for *D-F*. Contour values for the maps below each fish are 0.32, 0.64, 0.96, and 1.28 mV and correspond to 20, 40, 60, and 80% of the peak ΔV caused by the smallest object. The organ discharge of this apteronotid is of the high frequency, AC wave type (700-900 Hz); see Fig. 3.19 (Bastian 1986)

however, neither behavioral context nor adaptive significance of the system are understood very well.

Both conducting and non-conducting objects distort a fish's electric field, but the effects are of opposite nature and are discriminated by the fish. A good conductor compresses the isopotentials in the space that separates it from the fish's skin, thereby increasing the transepidermal voltage that stimulates its electroreceptors (Fig. 2.22B). An insulator, on the other hand, decreases the density of isopotentials; hence, the transepidermal voltage gradient. The difference field between disturbed and undisturbed fields shows this effect most clearly (Fig. 2.22C).

An object that differs in conductivity from that of water will cast an "electrical image" on the fish's skin, as represented by a change in transepidermal voltage (ΔV). ΔV can be measured by an electrode introduced in the gut, or implanted in the musculature, and an electrode placed on the skin (Fig. 2.22D-F).

Depending on the species, it is primarily or perhaps sometimes exclusively the tuberous, not ampullary electroreceptors that respond to the fish's electrolocation signal, the organ discharge (see Sect. 2.1). In mormyrids only one type of tuberous electroreceptors, the mormyromasts, mediate active electrolocation, afferent input from the Knollenorgane being blocked by a corollary discharge that is associated with the organ command. The reafference from ampullary organs that was discovered only relatively recently, is also blanked by a "central expectation" signal that is triggered by the corollary discharge (see Fig. 2.13).

In gymnotiform wave fishes, tuberous electroreceptors of the P and T types are distinguished, and of the B and M types in pulse fishes. P and B types code for stimulus amplitude and are certainly involved in active electrolocation. T and M types mark the timing of an organ discharge and probably mediate the fishes' sensitivity to capacitive shunts; thus, both types of tuberous electroreceptors would be involved in active electrolocation in gymnotiforms. Bastian (1987a,b) has found evidence for a small but significant supplementary role for the afferent input from ampullary electroreceptors in *Apteronotus leptorhynchus*.

The change in transepidermal voltage, ΔV, is determined mainly by the distance between an object and the fish's skin (given that the object's conductivity differs sufficiently from that of water). ΔV decreases as approximately the negative second power of the object's distance; thus, doubling of the distance results in reduction of the amplitude of ΔV to one-quarter. Given a threshold of 20 µV, a 12-mm cylinder causes a measurable response out to an object-fish distance of 30 mm. At a distance of 2.5 cm, *Gymnarchus* detected a glass rod at a threshold diameter of 2 mm (Lissmann and Machin 1958). Subsequent behavioral data also showed a detection limit of glass or plastic rods 2-4 mm in diameter at distances up to 2-4 cm from the fish. Within limits, an object must increase ninefold in diameter to compensate for a doubling of its distance (Bastian 1986). Active electrolocation thus is limited to very short ranges.

Weakly electric fish show extensive exploratory behavior when presented with a new object in their environment. These have been described as various "probing motor acts" in mormyrids (Belbenoit 1970; Toerring and Belbenoit 1979; Toerring and Moller 1984), but are also observed in gymnotiforms. They may consist in the animal's swimming to and fro past the object with a lateral distance of 1 or a few cm. From a distance, the fish often approaches an object by swimming backwards. Knife fishes like *Eigenmannia* often bend their tail around the object. These behaviors are thought to support electrolocation, by adding temporal and spatial cues to the sensory representation of the impedance inhomogeneity that is the target. The discharge is amplitude-modulated by these movements (as seen by a specific receptor); peripheral and central neurons respond to these low-frequency AMs (Bastian 1986).

Although the range of active electrolocation with a weak organ is so limited by physical constraints, fish seem to profit from its use. For example, *Gymnarchus* is a very successful hunter of other fish (although it is difficult to decide whether it uses electrical rather than mechanical cues. It certainly has poor vision and no barbels to detect its prey at a distance). *Gymnarchus* will respond by a sudden movement when a permanent magnet is moved past its body in air (Lissmann 1958; pers. observ.). Bastian's (1987a) experiments with *A. leptorhynchus* suggest that the active electrolocation system detects a moving target at a greater distance than does the ordinary mechanoreceptive system. Only the electric system allows fish to quickly classify an approaching object as to whether it may be "friend" (insulator, no risk) or potential foe (conductor). All fish that I have tested show a distinct dislike for metal objects.

Fish also use their electrolocation abilities to maintain a certain distance and posture relative to the substrate (Meyer et al. 1976; Feng 1977) and boundaries. This may be one of the main functions of active electrolocation in many weakly electric fishes that do not hunt large and quick prey (for example, the planktivorous *Rhabdolichops*; Lundberg et al. 1987).

Expanded into an ecological context, the above observation might indicate an adaptive advantage of prime importance that is associated with active electrolocation. Riverine species that are active at night, leaving their sheltered crevices to enter fast-flowing currents, as many weakly electric fishes do, need a sensory reference mechanism relative to ground in order not to be swept away. In swimming close to ground or rooted vegetation, these fishes could monitor and direct the nocturnal migrations some of them are known to perform (see Sect. 3.2.2.1). It is interesting to note that the other large group of nocturnal teleosts, catfishes, are also electroreceptive (although usually without electric organs, excepting *Malapterurus*, and the two *Synodontis* species of Hagedorn and Finger, unpubl.; see Sect. 3.1.3).

Another important function may be the location of social partners [mates, rivals, young fry that is guarded by the male in at least one mormyrid species (see Sect. 4.1.1.1), or eggs that are carried to the nest]. For example, during the elaborate courtship displays in *Pollimyrus isidori*, mates pirouette around each other (horizontally and also vertically at a later stage), precisely synchronized in the dark (Fig. 4.14).

2.3.2 Electrocommunication: Spatial Aspects of Sending and Receiving Electric Organ Discharges

While the useful range of active electrolocation is restricted to a few centimeters, the communication range at which a weakly electric teleost may detect another as a dipole source is far greater.

Knudsen (1975) has studied the spatial aspects of the electric field generated by gymnotiform fish (Fig. 2.21). He concluded that the field potential V falls off with the inverse square of the distance, and the field potential gradient E (a vector) with the inverse cube of the distance. These conclusions follow from the dipole equations for a volume conductor (as modified from electrostatics theory, see any general physics textbook, by Knudsen 1975):

$$V = \frac{\varrho_0 I}{4\pi} \frac{P_{\mathrm{I}} \cdot \cos \Theta}{d^2},$$

where ϱ_0, resistivity [$\Omega \cdot m$];
I, current [ampere]; Θ, the angle subtended by a ray from the midpoint of the dipole to a designated point, at a distance d (that is much greater than s, the separation of the current source from the sink); P_{I} represents $I \cdot s$, the 'dipole moment' [ampere $\cdot$ meter].

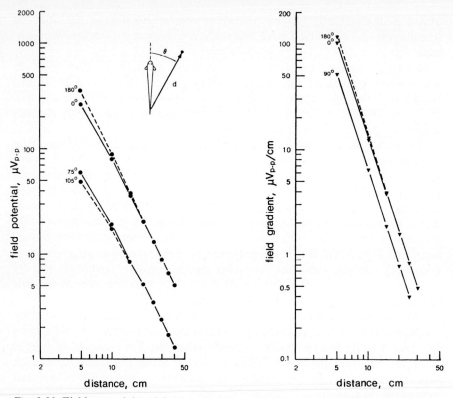

Fig. 2.23. Field potential and field gradient as a function of the distance d for a 6-cm *Apteronotus albifrons* in 3.8 kΩ · cm water (or 263 μS · cm^{-1}). The angle Θ defines the rays along which potentials were measured (see *inset*). As predicted by a dipole field, the field potential of this fish falls off with the square of the distance, and its field gradient with the cube of the distance (Knudsen 1975)

V depends on the angle Θ (see Fig. 2.23) and increases with decreasing water conductivity. The latter effect has been measured both in gymnotiforms (Knudsen 1975) and mormyrids (Bell et al. 1976; Squire and Moller 1982). In *Gnathonemus petersii*, for example, the peak-to-peak voltage of its biphasic organ discharge increased with resistivity up to 30 kΩ · cm (or 33 μS · cm^{-1}). It remained roughly constant to resistivities of 180 kΩ · cm (5 μS · cm^{-1}) due first to summation of an increasing first phase (behaving in battery-like fashion) and decreasing second phase; and second, to the low resistivity attenuation (Bell et al. 1976).

The field potential gradient, E, in a volume conductor is given by (see Knudsen 1975):

$$E_{(0°,\,180°)} = \frac{\varrho_0}{4\pi}\frac{2P_\mathrm{I}}{d^3}$$

$$E_{(90°)} = \frac{\varrho_0}{4\pi}\frac{P_\mathrm{I}}{d^3}.$$

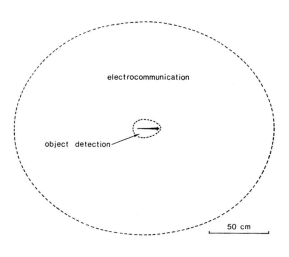

Fig. 2.24. The range of electrocommunication as compared with that of active electrolocation in a 18.6-cm *Eigenmannia* in 2 kΩ · cm water (or 500 μS · cm^{-1}). The range of electrocommunication is inferred from field measurements and sensitivity data; that of electrolocation is the threshold distance for a 2-mm plexiglass rod (Knudsen 1975)

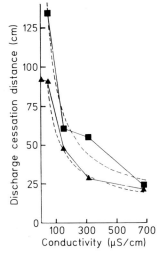

Fig. 2.25. Threshold distance of discharge cessation as a function of water conductivity and shelter tube orientation in the mormyrid, *Brienomyrus niger* (*triangles* end-to-end; *squares* parallel orientation). *Dashed curves* are best fitting lines determined by least-squares linear regression. Note steep increase of electrocommunication range with decreasing water conductivity (Squire and Moller 1982)

Knudsen has shown that these relationships hold true for weakly electric fish, except in the near field. The larger the fish, the more asymmetric is the head part of its field. The dipole moment increases dramatically with a fish's size.

The calculated communication range of a fish is of elliptic shape (Fig. 2.24). Its range depends on water conductivity for two reasons: in water of higher resistivity, the dipole moment increases, while the sensitivity of the fish for electrical stimuli decreases. High frequency sensitivity increases as a power function with decreasing water resistivity, down to 1 kΩ · cm (Knudsen 1974). The net effect is an increase of a fish's communication range in water of higher resistivity (see below).

Electrocommunication distance was determined in a small mormyrid, *Brienomyrus niger* (Squire and Moller 1982). At each end of a tank 4 m long (0.7 m wide, filled to 0.4 m) a fish was confined to a porous pot shelter, one of which was slowly moved towards the other. At a natural water conductivity of 52 µS · cm^{-1} (a resistivity of 19.2 kΩ · cm) one fish clearly detected the other at a distance of 135 cm, as shown by a long discharge cessation. The associated field gradient experienced by the responsive fish was 0.02 µV/cm. When one fish was moved back again towards its starting point a discharge rebound occurred at an even greater inter-fish distance of 157 cm (a field gradient of 0.01 µV/cm). Increasing water conductivity dramatically lowered these threshold distances. At an unnaturally high water conductivity of 678 µS · cm^{-1} (or 1.5 kΩ · cm), the distance at which discharge cessations occurred shrank to a mere 22 cm (a field gradient of 0.36 µV/cm; Fig. 2.25).

The smaller fish in a pair was more likely to "stop" than the larger (mean standard length: 11 ± 1 cm). Compared to Fig. 2.24 the elliptic communication range appeared rotated by 90° relative to the fish (but this may be due to compression of the fish's field by the long and narrow tank).

As known from many playback experiments in mormyrids and gymnotiforms (see Chap. 4), a fish will find an electric dipole source. Schluger and Hopkins (1987) and Davis and Hopkins (1987), who studied the fishes' approach path, conclude that the fish follows the current lines (field gradients) of a dipole field, reaching the source by a curved path. As may be expected from the physics, and observed by many scientists, the waveform of electric organ discharge does not change with distance and angle (except polarity reversal), nor are there conduction time differences or echoes (as in the acoustic modality) that may be exploited for locating a dipole source (see Hopkins 1986b for a comparative discussion). A recipient fish does not seem to have any idea of the distance or location of an electrically active target when it starts its approach, which resembles a "hunting" strategy (klinotaxis; Fraenkel and Gunn 1940, reviewed by Tinbergen 1979; Schöne 1980). However, a fish seemed disoriented when stimulated by a dipole at the periphery of the tank, the axis of which pointed at the fish's starting position at the center of the circular test arena (Schluger and Hopkins 1987; Davis and Hopkins 1987). This may be due to distortion of the dipole field by the wall of the test tank.

Following the lines of force by a klinotaxis strategy resembles the galvanotaxis behavior of the unicellular protozoan *Paramecium* in a DC field (Verworn 1889; review Machemer 1988), except that fish respond to much weaker AC or pulsed field gradients and do not always approach the same electrode (the cathode). A slight, but statistically significant, preference for the "electrical head" of a stimulus dipole simulating (although only crudely) conspecific EOD pulses was, however, observed in *Gymnotus carapo* by Davis and Hopkins (1987), confirming Westby (1974).

In sharks and rays, Kalmijn (1988) summarizes another approach algorithm to dipole sources that guides the fish correctly to the target under all angles, confirmed by extensive observation.

Chapter 3

Species Diversity of Electric Organ Discharge Activity

The discharge activity of an electric fish depends on the complex anatomical and physiological detail of its electric organs (for the discharge waveform) and central nervous command structures (for the discharge rate). Phylogenetic groups and, within a group, the species it comprises, differ in these structures and their physiological properties (see Chap. 2); these differences are reflected in a wide variety of EOD activities.

The EOD of a weakly electric fish is one of the best examples of spontaneous behaviors of all neuroethology (for circadian rhythms, see Sects. 3.2.1.1 and 3.2.2.1). At least resting discharge activities, but probably several other patterns as well, are innate behaviors par excellence, as may also be concluded from developmental studies (reviewed in Kirschbaum 1984).

Therefore, at least some properties of the EOD activity reflect phylogenetic relationships as well as adaptive specializations, aiding in the identification of taxa or even species, once a catalogue has been established (see Chap. 1). Just as with any other character, there are two major pitfalls to be avoided: those represented by (1) convergent evolution, and (2) a wide intraspecific variability of an EOD character, which may mistakenly lead scientists to consider a collection of specimens as representing two or more species (see Roberts and Stewart 1976), or to consider a character to be sexually dimorphic although there may be no more than a statistical trend (see Bratton and Kramer 1988). That a broad context is necessary in the study of systematics questions is also shown by Lundberg and Stager's (1985) discovery of considerable microgeographic diversity in a knife-fish. Therefore, if a variant EOD activity is used as a quick way to distinguish species (or the two sexes of a species), an independent criterion (or rather set of criteria) is necessary to establish the validity of the distinction.

The EOD activity of an undisturbed, isolated fish shows three aspects of species-characteristic properties: (1) the waveform of the discharge pulse or wave (that is, voltage change over time); (2) the pattern of temporal spacing of discharges; (3) discharge rate responses to external disturbances.

3.1 Waveforms of Electric Organ Discharges

Electric teleosts are usually divided into wave and pulse species. When played back by loudspeaker, wave and pulse EODs sound characteristically different to the human ear: wave EODs resemble the steady tone of a musical instru-

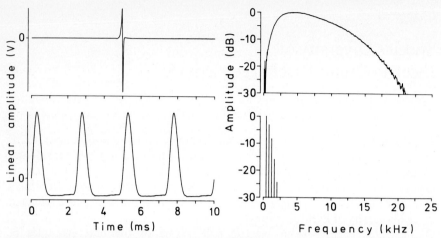

Fig. 3.1. Pulse and wave discharges. *Left* Oscillograms of the EODs (head-positivity is upwards); *right* amplitude spectra with the amplitudes expressed as dB attenuation relative to the strongest spectral component. Same time and frequency axes. Note that EOD-pulses like that of the African mormyrid, *Gnathonemus petersii*, usually are very short and show a broad-band spectral frequency content. They are separated by long intervals. These signals are click-like and "noisy". Wave-EODs, like that of the South American gymnotiform, *Eigenmannia*, are continuous, and the energy is concentrated at discrete frequencies or spectral lines: the fundamental frequency (or first harmonic which often is the strongest component, as found here), and its higher harmonics or overtones at integer multiples of the fundamental frequency. In any periodic signal the fundamental frequency is the repetition frequency of a full period. Such a signal resembles the steady tone played on a musical instrument; it is harmonic

ment such as a flute or a violin, with characteristic differences among species, or the sexes of a species, in timbre or frequency (for example, Figs. 3.14 or 3.15). Wave EODs sometimes resemble a sinusoid (Fig. 3.14) but can be amazingly complex in form, and rich in harmonic content, in other species (Fig. 3.19). Frequencies range from about 50 to about 1800 Hz.

Pulse EODs sound like clicks and are emitted at a regular or a seemingly irregular sequence, at low mean rates (from below 1 Hz to about 65 Hz at rest). When played back by an audio monitor they sound like the action potentials of a spontaneously active nerve cell.

Because the pauses between pulse discharges are long (and often variable) compared with the duration of the discharges, a single discharge is considered the signal and may be analyzed for its spectral contents (Fig. 3.1).

There is a broad and continuous spectrum with a flat peak region; that is, the signal is "noisy" (geräuschhaft). This contrasts with the harmonic "tone" signal emitted by a wave species like the South American knife-fish *Eigenmannia*: usually the EOD frequency is extremely stable (the discharge is periodic), so that a whole series of discharges can be analyzed for its harmonic content. This yields line spectra in which the energy is concentrated at discrete frequencies: the fundamental frequency (which is the repetition frequency of a full period of the signal) and its integer multiples, the higher harmonics or overtones.

When electric organs evolved in the ancestors of Mormyriformes and Gymnotiformes, strong selection pressures must have shaped their EODs as either pulse or wave. At present, we can only advance speculations about the nature of these selection pressures. The gymnotiform *Distocyclus goachira* (Fig. 3.18), with its intermediate discharge, may well be unique.

The 188 species of the Mormyridae (see Chap. 1) seem to be all pulse species, with the monotypic *Gymnarchus niloticus* of the related family Gymnarchidae representing the only African wave species. The South American Gymnotiformes also comprise both pulse and wave EOD families; however, wave species outnumber pulse species. Most Mormyriformes and Gymnotiformes seem to feed mainly on insect larvae, with a few piscivorous species among the larger ones. Some species are gregarious, some solitary, both of the wave and the pulse EOD type. All seem to be more active at night, with some species almost totally inactive during the day. At present we are unable to discern a clear pattern of adaptive strategies of weak electric organ discharges (being either pulse or wave), as related to feeding habits or ecology. The character pulse or wave EOD is, however, strongly linked to taxonomic groups (Chap. 1).

An advantage pulse EODs might have over wave EODs is in electrolocation. Because of their wide and often high spectral frequency content (Figs. 3.1, 3.7), pulse species may detect natural capacitances (such as leaves or food items) of a wide range as impedance inhomogeneities in their environment (see Meyer 1982). Most wave species tend to concentrate the energy at only one frequency which is, in addition, comparatively low (below 1 kHz); they should be less sensitive. However, some apteronotids such as *Sternarchella* sp. 1 (Fig. 3.19) distribute the energy over several harmonics, with a high harmonic (in *Sternarchella's* EOD the fifth) being the strongest; these wave species should rival pulse species in their sensitivity for capacitive impedances.

For a communication function, wave EODs might be at a disadvantage. To signal different messages their modulation is limited to relatively moderate frequency increases and decreases (apart from going silent), whereas pulse fish, such as the Mormyridae, may encode messages in complex patterns of inter-pulse intervals, in addition to often dramatic changes of their pulse rates (see Chap. 4).

Compared with pulse EODs, frequency resolution is incredibly high in wave EODs, because a wave fish can determine the EOD frequency of another fish as the difference frequency from its own frequency, by beat frequency analysis of the superposition signal (more familiar from mistuned musical instruments playing almost the same note). Therefore, it is not surprising that in a few wave species at least, the two sexes discharge at distinctly different frequencies. Another way of encoding the sex of a sender is the variation of the form of a wave signal along with its harmonic content; this is also seen in some wave fish.

In pulse species, intraspecific variability of the EOD may also be high. This concerns the amplitude, duration, or waveform of a pulse EOD. In some species the means of certain waveform parameters are statistically significantly different between the two sexes; in other species there are no such differences.

Wave EODs represent a continuous drain of energy for the sender; there are no strong electric wave fish. Compared with most wave EODs, pulse EODs are of lower repetition rate and stronger amplitude (the EODs of the mormyrid *Mormyrus rume* are so strong they are felt by the human hand touching the fish's tail, even in a less than half-grown individual of 20.5 cm total length; B. Kramer, pers. observ.). Pulse EODs might be detected over a greater distance because of their strong amplitude. This would be an advantage both for communication and active electrolocation. However, wave EODs compensate for being weak by strongly contrasting from background noise by their harmonic structure. There is no or little DC component in the EOD of wave fishes (only a few studied) unlike that of many pulse species.

An important source of noise are conspecifics and other electric fish, especially those with discharges of wideband frequency content. In a pulse species, spectral frequency analysis should therefore be of little use in gaining information about other individuals of a group. The members of a group of a wave species, however, discharge at different frequencies within a usually wide species-specific frequency range. Should the frequency difference between two wave fish become too small, one or both may show the so-called jamming avoidance response (JAR), usually a frequency shift increasing the difference. To spread out individual frequencies for better individual recognition could be the biological main function of the jamming avoidance response, at least in *Eigenmannia* (see Sect. 4.2.2.2). Frequency analysis thus could inform a wave fish about the number and, at least in some species, "quality" (such as sex or age) of other members of its group.

The sensory capacity for discriminating sexually dimorphic wave EODs with fundamental frequency or intensity cues not being factors, but with characteristic differences in waveform and harmonic content, has been demonstrated in *Eigenmannia* (see Sect. 4.2.2.2).

3.1.1 Mormyriformes

The African Mormyriformes comprise two families, the Mormyridae and the monotypic Gymnarchidae. The Mormyridae discharge their electric organ in short pulses, while *Gymnarchus niloticus* is the only known African wave species.

3.1.1.1 Mormyriform Pulse Species (Mormyridae)

The Mormyridae are a large family of 18 genera and 188 species (subspecies excluded; Gosse 1984). Their total lengths vary from small (up to 90 mm, such as *Pollimyrus isidori*) to very large (1.50 m, such as *Mormyrops deliciosus*). Most Mormyridae are still unstudied regarding their EOD activity. Especially

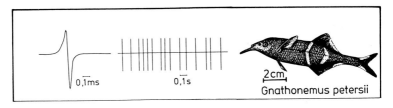

Fig. 3.2. Single EOD (*left* digitized at 100 kHz), and train of EODs (*right*) of *Gnathonemus petersii*. Note different time *bars* (factor of 1000). In a pulse fish, the intervals between EODs are long compared with the duration of an EOD (Kramer 1985b)

some of the larger species appear to be known only from fish markets (as cited in Gosse 1984). There are as yet no non-electric species.

The EODs of Mormyridae, as known at present, are pulses which are short compared with the long and variable intervals which separate them (Fig. 3.2).

We do not know a single wave species. The EOD pulse is typically less than 1 ms but may be shorter than 250 µs in several small species, such as *P. isidori*, with the main spike around 50 µs or even less (Fig. 3.4). These EODs resemble extracellularly recorded nerve action potentials with characteristic differences in waveform: they are nearly monophasic (for example, *Mormyrus rume*), biphasic (example: *Gnathonemus petersii*; Fig. 3.1), or triphasic (example: *P. isidori*; Fig. 3.4). The EODs are sometimes preceded or followed by small additional potentials; the intraspecific variability is, however, largely unstudied (excepting a few species like *P. isidori* and *G. petersii*; see below).

The waveform of an EOD pulse of an individual *G. petersii*, *P. isidori* or *Petrocephalus bovei* is extremely stable over short periods of time (hours and days) but does show some plasticity over weeks and months under constant conditions (Kramer and Westby 1985; Bratton and Kramer 1988). The EOD duration does not change much with temperature; the Q_{10} is only 1.5 in *G. petersii* (Kramer and Westby 1985). Compared with the temporal stability of the EOD within an individual *G. petersii* the variability between individuals is great (Fig. 3.3).

Sexually dimorphic EODs have been suggested in several mormyrid species (review: Hopkins 1986a). However, these observations need further support, as taxonomical and other questions are still open, and data on intraspecific variability exceedingly scarce (see Bratton and Kramer 1988, also for dependence of EOD waveform on water conductivity; see also below). According to Hopkins and Bass (1981) a mormyrid called *Brienomyrus brachyistius (triphasic)* discriminated between male and female EOD waveforms in playback experiments in the field at night. These authors have also proposed an interesting sensory mechanism of temporal EOD waveform discrimination. The discrimination capacity would be better supported by laboratory studies under more explicit experimental control (see Kramer 1985a, p. 63-64; Kramer and Weymann 1987, concerning playback of tape-recorded EODs).

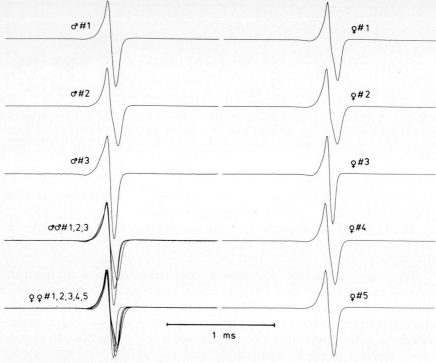

Fig. 3.3. Individual variability of EOD waveforms in the mormyrid *Gnathomus petersii*, normalized to the same head-positive amplitude, and centered on the time of zero-crossing. The two figures at *lower left* show discharges of males and females superimposed. These examples are representative of a group of 27 fish in which no sex difference could be discerned. Preserved fish are stored in the Zoologische Staatssammlung, München (FRG), Nos. ZSM 27167-27169 (Kramer and Westby 1985)

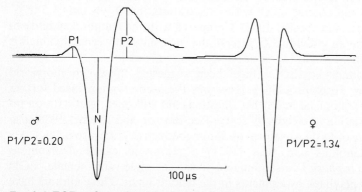

Fig. 3.4. EODs of a male and a female *Pollimyrus isidori*, normalized to the same N-wave height, showing a marked difference in their ratios of P1/P2-amplitudes. P1 is the first head-positive phase, P2 the second. Individuals selected for demonstrating a presumed sexual dimorphism now known to represent a statistical trend only within a wide intraspecific variability. Conductivity, 100 $\mu S \cdot cm^{-1}$; 2 MHz digitization (Bratton and Kramer 1988)

Table 3.1. EOD waveform variability at 100 µS · cm^{-1} in *Pollimyrus isidori* (as ranges and means ± standard deviation)[a]

	Males ($n=10$)	Females ($n=14$)	Mann-Whitney U-test
P1/P2 ratio	0.04 – 0.94	0.37 – 3.33	
Mean P1/P2 ratio[b]	0.49 ± 0.26	1.17 ± 0.82	30.5
Mean 100 (P1–P2)/N[b]	−19 ± 13	−4 ± 13	31.0
N duration (µs)	23.1 – 37.8	17.3 – 31.6	
Mean N duration[c]	28.6 ± 5.3	25.8 ± 3.9	52.0
P1–N separation (µs)	17.8 – 28.9	15.8 – 25.8	
Mean P1–N separation[c]	23.0 ± 3.6	20.5 ± 2.7	45.0
P1–P2 separation (µs)	33.8 – 60.9	28.9 – 52.4	
Mean P1–P2 separation[c]	47.0 ± 9.5	41.8 ± 7.4	51.0
Peak amplitude frequency (kHz)	8.0 – 20.0	10.5 – 25.0	
Mean peak amplitude frequency (kHz)[c]	13.4 ± 4.0	16.4 ± 4.4	42.5

[a] Definition of phases, see Fig. 3.4. N-wave duration measured as the time between zero-crossings. Peak amplitude frequencies determined from amplitude spectra (as in Fig. 3.7). Note that there is a statistically significant difference of the mean P-ratio between the sexes (second line), although the female range of P-ratio overlaps two-thirds of the male range (first line). All other waveform parameters do not differ significantly between the sexes (Bratton and Kramer 1988)
[b] Differences significant at $P < 0.025$ (Mann-Whitney U-test, two-tailed).
[c] Difference not significant ($P > 0.10$).

There are no separate female and male ranges in the EOD waveform variabilities of *G. petersii* (Kramer and Westby 1985). The EOD waveforms of *P. isidori* vary individually (Lücker and Kramer 1981); in contrast to *G. petersii* there is a statistically significant difference between the two sexes in the means of one waveform parameter, the P-ratio (Westby and Kirschbaum 1982; Bratton and Kramer 1988), and no difference in several others (Fig. 3.4; Table 3.1).

A distinct sexual dimorphism is, however, not present because of the wide overlapping of the distributions (Table 3.1). (For a discussion of statistical distributions of sexually dimorphic and homomorphic characters, see Burkhardt and de la Motte 1985, 1987, 1988).

One reason for the presence of a sexual dimorphism, or an only statistical character difference between the sexes, is sexual selection (such as female choice; for a review, see Wilson 1975). In *P. isidori*, however, physiological reasons can explain the observed facts more parsimoniously (Bratton and Kramer 1988):

Androgen hormones (such as testosterone) are well-known for their morphogenetic and anabolic effects; that is, the stimulation of growth and strength of skeletal muscles in vertebrates (see, for example, Eckert et al. 1988; Blüm 1986). This has also been shown in the electric organs (which are derived from skeletal muscle) of androgen-treated female mormyrids (review: Bass 1986, p. 52; Landsman and Moller 1988), and in a gymnotiform pulse fish (Hagedorn and Carr 1985). Examples of a stronger EOD amplitude in males compared with females (both untreated) are the gymnotiforms *Eigenmannia lineata* (Kramer 1985a) and *Hypopomus occidentalis* (Hagedorn in prep., cited from Hagedorn and Carr 1985).

Under the influence of endogenous androgens, male *P. isidori* should develop a stronger electric organ, yielding, on average, a stronger first phase (P-wave) current than the electric organs of females. (The duration of the P-wave is about twice that of the N-wave in *P. isidori*.) Therefore, the electrically evoked second phase (N-wave) should occur at a shorter latency relative to the onset of the P-wave in males compared with females. The result would be a smaller P-ratio in males than in females [see the numerical model by Westby (1984) who simulated EODs of all P-ratios by appropriately delaying the N-wave relative to the P-wave].

The electric current produced by the electric organ is subject to an external resistive load, which varies widely in natural African habitats because of fluctuating water conductivities (Gosse 1963; Bénech and Quensière 1983). The resistive load affects both first- and second-phase current of *G. petersii*'s EOD (Bell et al. 1976), and the range of communication (Squire and Moller 1982).

In both *P. isidori* and *P. bovei*, and surely other species as well, conductivity seriously affects the waveform of the EOD, especially in the ecologically most relevant conductivity range of 5 to 110 $\mu S \cdot cm^{-1}$ (Fig. 3.5).

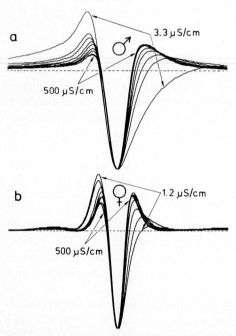

Fig. 3.5. EODs of a male (a) and a female (b) *Pollimyrus isidori* recorded through a range of water conductivities ($\leq$ 3.3, 5, 10, 20, 50, 100, 200, 500 $\mu S \cdot cm^{-1}$), normalized and superimposed. Note increase of the first head-positive wave (*P1*), and decrease of the second (*P2*), in both waveforms with decreasing conductivity (Bratton and Kramer 1988)

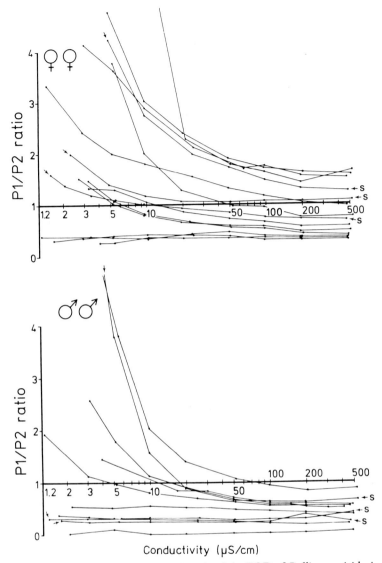

Fig. 3.6. The effect of water conductivity on the P-ratio of the EOD of *Pollimyrus isidori* (P-ratio: see Fig. 3.4). *Top* 14 females; *below* 10 males. Note that the P-ratio changes most in the ecologically relevant range (5-110 µS · cm^{-1}), and that the P-ratios in some individuals of both sexes are not affected by conductivity changes. Thus, the intraspecific variability observed at a relatively high conductivity (100 µS · cm^{-1}; Table 3.1) is higher still at lower conductivities. Three spawning males and three spawning females are indicated by *arrows S* (Bratton and Kramer 1988)

With decreasing conductivity the P1 wave increases relative to the P2 wave in most animals tested, irrespective of sex (Fig. 3.6).

This effect augments the overlapping of the P-ratios of the two sexes observed at the same conductivity (see Table 3.1). With decreasing conductivity the N-wave broadens, because second phase current depends on first phase

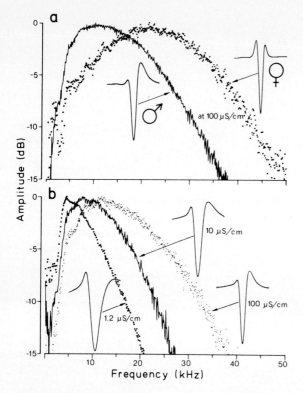

Fig. 3.7. Fourier amplitude spectra of *Pollimyrus isidori* EODs. *Ordinate* gives the amplitude as dB attenuation relative to strongest spectral component; *abscissa* frequency (kHz). *a* The two fish were selected for a great difference in P-ratio (see waveform *insets*). The female's discharge peaks at a higher frequency and shows a greater band-width at half power (−6 dB) than the male's discharge. *b* With decreasing conductivity the head-negative wave of an EOD widens (see *insets*), and the spectral peak frequency declines (in this case from about 12 to about 4 kHz; Bratton and Kramer 1988)

current (Bell et al. 1976). The broadening caused the spectral peak frequencies to decline dramatically (Fig. 3.7).

Because of the wide overlapping of all EOD characters of the two sexes at the same conductivity, and their strong dependence on a seasonally and regionally variable environmental factor, the waveform or the spectral features of an EOD pulse do not appear to be reliable criteria for mate recognition in *P. isidori*. However, the differences between the EODs of different species may be so great that they probably are resolved. The durations of EODs of different mormyrid species vary by a factor of about 100 [from below 0.1 ms (Fig. 3.4) to at least 8 ms (Hopkins 1986a)].

Knollenorgan electroreceptors are the putative communication receptors in mormyrids, though they have not yet been studied in detail. From preliminary data (Hopkins 1983) it appears that they are exceptionally weakly tuned; thus it is unlikely they could resolve other than very marked differences in spectral frequency composition of EODs, especially as these have broadband characteristics (the half-power bandwidth in *P. isidori*'s EODs often exceeds 20 kHz; see Fig. 3.7).

In the context of communication, the function of the mormyrid electric organ appears to provide impulse-like, precise time-marks. These are essential in a communication system encoding information in time intervals (see Chap. 4, especially Sects. 4.2.1.2 to 4.2.1.4). The shorter the impulse the better the time resolution of the interval; this may be one of the key selection pressures which made many mormyrid electric organ discharges much shorter than nerve action potentials, some of them probably representing the fastest electric phenomena in all biology.

The time-marker function would be largely independent of changes in the environment (temperature, conductivity). Whether the fish can access additional information contained in the EOD waveform, which is, in a few species at least, subject to modification by the changing environment (mainly conductivity), remains as yet unanswered. There are well-documented alternative mechanisms of information exchange in mormyrids (see chapter 4).

3.1.1.2 Mormyriform Wave Species (Gymnarchus)

The only mormyriform wave species known is the monotypic *Gymnarchus niloticus* Cuvier 1829 (family Gymnarchidae) which grows up to 1.51 m (cited from Gosse 1984), or 1.60 m (Svensson 1933; cited from Lissmann 1958). *Gymnarchus* has a swim bladder adapted for atmospheric respiration (with a modified circulatory system). The fish gulps air at the surface, often near fallen trees in the river. It preys on other fish. Prey appears to be sucked in by an expansion of the branchial basket and is accompanied by a resounding snap (Lissmann 1958).

At rest, *Gymnarchus*' weak electric organ discharge of about 220-400 Hz (Fig. 3.8) resembles that of the gymnotiform *Eigenmannia* (Fig. 3.1). There are further aspects of convergent evolution (see Chap. 1).

As in *Eigenmannia*'s EOD, head-positive pulses are superimposed on a head-negative baseline; there is no or little DC component (Bennett 1971a).

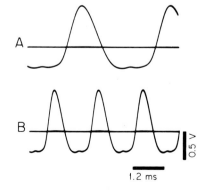

Fig. 3.8 A,B. Wave EOD of the unique African wave species, *Gymnarchus*. Animal was about 30 cm long. Zero potential level indicated by the *horizontal line*. *A* Expanded time scale (Bennett 1971a)

This is because the uninnervated posterior faces of the electrocytes of the electric organ are inexcitable and pass current only capacitatively.

In three specimens of 38, 42, and 52 cm, the maximum voltage of the electric organ was 3, 7, and 4 V, respectively (measured in air with copper electrodes placed at the tip of the tail and at one-third of a fish's body length from the snout, near the rostral end of the electric organ; fish were superficially dried). No shocks were felt by placing wet fingers on a fish's tail or other parts of its body (Lissmann 1958). This is unlike the smaller *Mormyrus rume* which, although "weakly electric", like, apparently, all other members of the family Mormyridae, gives off weak but distinct shocks felt by fingers placed on its caudal peduncle (see Chap. 3.1).

3.1.2 Gymnotiformes

The South American Gymnotiformes comprise six families. In contrast to the Mormyriformes, of which all but one are pulse species, wave species outnumber pulse species in the Gymnotiformes. One family, represented by the monotypic electric eel, is both weak and strong electric. All Gymnotiformes have an elongate, often compressed body (hence the name "knife-fishes"), no (or very small) tail fin, no (or only a rudimentary) dorsal fin, and move by an undulating anal fin (the unique mormyriform *Gymnarchus* propels itself in a similar fashion by its undulating *dorsal* fin). For the hydromechanical properties of the gymnotiform mode of swimming, see Blake (1983a,b).

Knöppel (1970) studied the gut contents of six species of gymnotiforms, of the families Gymnotidae, Sternopygidae and Rhamphichthyidae, from small forest streams of the terra firme with little seasonal variation. "The species ... present eat larvae of insects, and plant matter was always found. There were percentages of fish and Crustacea in *Gymnotus carapo*; in *Gymnotus anguillaris* there were percentages of ants."

There is no useful key to the Gymnotiformes; the recent classification by Mago-Leccia (1978) has the advantage of grouping pulse and wave-species into separate families of only one discharge type (see below).

3.1.2.1 Gymnotiform Pulse Species

Gymnotiforms of four families discharge their electric organs in pulses: the Rhamphichthyidae, Hypopomidae, Gymnotidae, and Electrophoridae. EOD rates vary from less than 1 Hz to about 65 Hz at rest.

In a sympatric community of Gymnotiformes from the Solimoes (Amazonas) river, just above the confluence of the Rio Negro (near Manaus, Brazil), there were 11 pulse species with at least one representative from each family (Kramer et al. 1981a). The EOD waveforms showed characteristic differences among most species, and their durations ranged from 1 to 4.4 ms (Fig. 3.9).

Fig. 3.9. Waveforms and amplitude spectra of gymnotiform pulse EODs from a sympatric fish community near Manaus (Solimoes). *Left* EOD waveforms; each trace represents 10 ms. *Right* The associated amplitude spectra, with the *ordinate* showing logarithmic amplitude in dB; the *abscissa* logarithmic frequency in kHz. *a* The electric eel, *Electrophorus electricus*; weak discharge. *b Hypopomus* sp. 1, with a similar monophasic EOD; *c Hypopomus* sp. 2 with a biphasic discharge; *d Hypopomus* sp. 3; *e Hypopomus* sp. 4; *f Steatogenys elegans*; *g Rhamphichthys* sp. 1; *h Rhamphichthys* sp. 3; *i Gymnotus carapo*. Digitized at 12 bit/100 kHz from a Nagra tape-recording (38.1 cm/s); negative after-potential in the recordings of pure DC pulses (*a, b*) is artifact. Provisional species names refer to Kramer et al. (1981a)

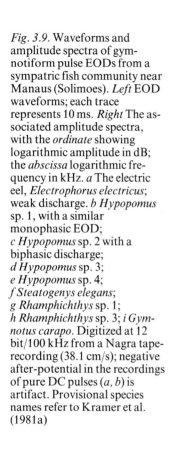
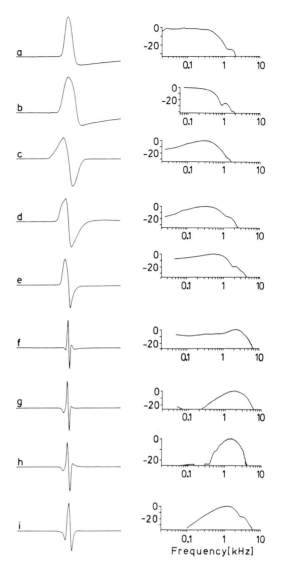

The monopolar, head-positive weak EODs of *Electrophorus* (2.1 ms) and those of *Hypopomus* sp. 1 (2.3 ms) were very similar and differed from all other species' EODs. These EODs had a flat amplitude spectrum from DC to about 200 Hz. The other species' EODs were essentially biphasic or triphasic, with additional pre- or post-potentials of weaker amplitude in some cases. The amplitude spectra of all of these EODs had well-pronounced peaks, with the energy concentrated in a certain frequency band which appeared narrow compared with the mormyrid spectra shown in the previous chapter. Spectral peaks of individual EODs ranged from about 300 Hz [*Hypopomus* sp. 2 (biphasic)] to about 2300 Hz (*Steatogenys elegans*). This is low compared with

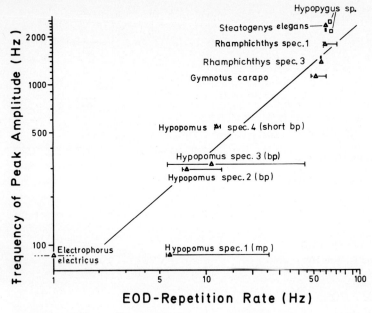

Fig. 3.10. Plot of EOD repetition rate vs frequency of peak amplitude, as found by Fourier analysis of single EODs, in sympatric pulse fish from the Solimoes (Amazonas). The *horizontal bars* refer to the limits of the pulse rate distributions, the *triangles* indicate the modes of these distributions (measured as intervals with a high resolution digital computer). A least squares regression line shows the positive correlation between the modes of pulse rate distributions and frequencies of peak amplitude. Each point represents one individual. Temperature varied between 27-29° C, water conductivity was 65-70 µS · cm^{-1} (Kramer et al. 1981a)

the Mormyridae in which spectral peaks of up to about 25,000 Hz are found (Fig. 3.7).

The frequencies of peak amplitude, as determined from Fourier amplitude spectra of single EODs, are significantly positively correlated with resting EOD rates across species ($r = 0.94$, $p < 0.01$); that is, pulse durations tend to decrease with shortening interpulse intervals. The frequency of peak amplitude of a species' EOD tends to increase with EOD repetition rate as a power of approximately 0.8; that is, with a 100% increase in EOD rate at rest, the peak amplitude frequency increases at a rate of 75-80% (Fig. 3.10).

It is doubtful whether any one species' EOD differs sufficiently from all other species' EODs in spectral peak frequency to allow species identification (in the sense of an intrinsic isolation mechanism), except the two monophasic species (*Hypopomus* sp. 1 and *Electrophorus*), the EODs of which are clearly lower in frequency content than those of all other known species. Doubts should be expressed, because with a better knowledge of intraspecific variability (including geographical variability; Schwassmann 1976), and with a more complete collection of species sympatric for any locality, interspecific overlap will surely increase beyond what is known to date. Heiligenberg and

Bastian (1980) also found wide overlap of spectral EOD peak frequencies among their six sympatric pulse species from the Rio Negro. The study of intraspecific, microgeographic diversity has barely begun; neither in gymnotiforms (but see Lundberg and Stager 1985) nor in mormyrids have the data been collected to discuss current models of speciation (for example, Endler 1977, Stanley 1979).

Rather than peak amplitude frequency of a species' EOD, a related parameter, spectral bandwidth, might play a role in species recognition (for example, the -10 dB bandwidth relative to the frequency of peak amplitude). Bandwidth might be sensed by a receiver's population of tuberous electroreceptors tuned at different "best" frequencies (Viancour 1979a; Bastian 1977). The available spectral amplitude data do not, however, support the notion that such a mechanism might be important for species recognition, because differences between species with similar spectral amplitude peaks are not striking (Kramer et al. 1981a).

Perhaps it is the differences between species-specific EOD waveforms of similar spectral frequency content upon which mate and species recognition relies, or else a combination of parameters. These propositions have yet to be explored, and the mechanisms to be elucidated (see Sects. 3.2.2.1 and 4.2.2.1).

In spite of the impressive variety of EOD waveforms found in Gymnotiform pulse species, some species' EODs resemble each other. In these fish, no study of intraspecific waveform variability and dependence on water conductivity is yet available, comparable to that in the mormyrid *Pollimyrus isidori* (Sect. 3.1.1.1).

Hagedorn and Carr (1985) have reported a sexual EOD waveform dimorphism in *Hypopomis occidentalis* (Fig. 1.6D) from Panama. Males are larger and have wider tails and stronger EODs of longer duration. Accordingly, the mean spectral peak frequency is lower in males than in females (826.2 ± 200.4 Hz, $n=17$, vs 984.4 ± 97.6 Hz, $n=19$). The considerable overlap might be age- or maturity-related, as there is a significant negative correlation between male tail width and peak power frequency.

Westby and Shepperd (1986) report a very similar sex difference in *Hypopomus beebei* from French Guiana. The intraspecific variability is unknown.

We do not yet know whether pulse gymnotids can discriminate between such species or sex differences in their EODs (however, in *H. occidentalis* stimulation experiments suggest they might; see Sect. 4.2.2.1). Westby and Shepperd (1986) and Hopkins and Westby (1986) discuss sensory mechanisms which could achieve this; see also Shumway and Zelick (1988). However, no study is yet available on the effect of conductivity changes on EOD waveform (see Sect. 3.1.1.1). A possible effect should depend on whether the uninnervated faces of the electrocytes are excitable (yielding biphasic EODs) or not (yielding monophasic EODs). Biphasic EODs are probably sensitive to conductivity changes, monophasic EODs are probably not.

An interesting case is *Gymnotus carapo*, with its triphasic EOD. The three phases are explained in terms of a specialized anatomy and physiology of the electric organ of this species (see Bennett 1971a): the first head-negative phase

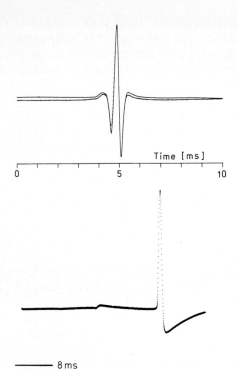

Fig. 3.11. Superimposed EOD waveforms of a *Gymnotus carapo* recorded at 96 μS · cm⁻¹ (*line*) and at 10 μS · cm⁻¹ (*points;* corresponding to *upper base line*), both at 26.8° C. EODs are normalized to the same height. Head-positivity is up. The animal had time to adapt to a new condition for a week. Note little effect of conductivity change. The experiment was repeated with another fish with similar result. On-line digitization at 12 bit/100 kHz

Fig. 3.12. "Neural command signal" (*center*) preceding electric organ discharge (*right*) in a pulse gymnotiform, the electric eel (weak discharge). Digitized at 12 bit/100 kHz from a Nagra tape-recording at 38.1 cm/s

is neurally evoked from a small part of the electric organ, the electrocytes of which are innervated anteriorly; 0.5 ms later the head-positive phase is also neurally evoked from the main part of the electric organ, with electrocytes which are innervated posteriorly (as usual). To this potential, which is the strongest phase of *G. carapo*'s EOD, the electrically excited second phase of the electrocytes first mentioned is added. Only the third phase (produced by the main part of the electric organ) appears exclusively electrically excited. Therefore, the first and the second phase should be completely or largely independent of conductivity changes, because both are exclusively or mainly neurally evoked. This independence is seen in Fig. 3.11.

In all four *Hypopomus* species and in *Electrophorus*, a weak signal precedes each EOD by approximately 13 ms (Fig. 3.12). This is probably a neural command signal, also seen, though at a shorter interval from the EOD, in the Mormyridae. A similar signal is seen in *Rhamphichthys* sp. 3, preceding an EOD by about 10 ms. *Rhamphichthys'* EOD repetition rate is considerably higher than that of the *Hypopomus* species mentioned.

The waveform of *Electrophorus'* strong discharge is similar to that of the weak EOD. A strong EOD volley is preceded by a remarkable postsynaptic summation potential consisting of about four phases and one weak EOD. The PSPs, the single weak EOD, and the ensuing volley of strong EODs are all at the same rate of 500 Hz or above. The single weak EOD may be absent if the interval from the last discharge (weak or strong) was shorter than about 60 ms (Bauer 1979; Fig. 3.13).

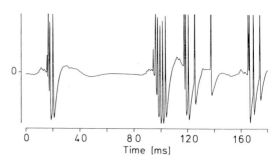

Fig. 3.13. Weak and strong EOD of the electric eel, *Electrophorus*, including PSPs. Note PSPs (of about 4 phases) followed by a single weak EOD, evoking a volley of strong EODs (*left* and *center*); the single weak EOD is absent when the interval from the last strong EOD is less than 60 ms (three instances on the *right*). Digitized at 12 bit/100 kHz from a Nagra tape-recording at 38.1 cm/s

These observations support Bauer's hypothesis of two independent commands in the electric eel, both acting through the single medullary relay nucleus (Bennett et al. 1964) and exciting the spinal electromotor neurons of all three electric organs with each command (Albe-Fessard and Chagas 1954). At low frequencies, only the electrocytes of the Sachs' organ are fired, while summation of PSPs would drive the remaining electrocytes above threshold only at high frequencies.

3.1.2.2 Gymnotiform Wave Species

Two gymnotiform families discharge their electric organs in a wave-like manner: the Sternopygidae and the Apteronotidae. EOD frequencies range from about 15 to about 1800 Hz. In a fish community near Manaus, 32 wave species were found (the species of lowest frequency may be considered intermediate between wave and pulse).

Sternopygids, which are considered the most primitive gymnotiforms (Mago-Leccia and Zaret 1978; Fink and Fink 1981), all have simple EOD waveforms, with the fundamental frequency being the strongest harmonic component, and higher harmonics declining in amplitude with frequency. Fundamental frequencies range from about 15 Hz to above 800 Hz (Fig. 3.14). An almost sinusoidal EOD has been observed in *Distocyclus conirostris*. Its EOD lacks a second harmonic; a few still higher harmonics are very weak (Fig. 3.14).

Young *Eigenmannia* (<10 cm) also show nearly sinusoidal EODs with weak second harmonics of around 10% of the amplitude of the fundamental frequency, or less (Fig. 3.15). The second harmonic increases in amplitude up to about 40% of the first harmonic in adult females, and up to about 70% in adult males (Fig. 3.16). The second harmonic deserves special attention, because (1) it is the strongest signal component next to the fundamental frequency, and (2) fish are still relatively sensitive at frequencies of around two

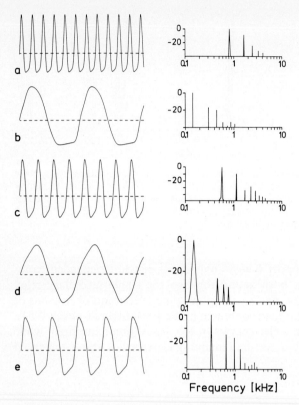

Fig. 3.14. Sternopygidae: waveforms and amplitude spectra of their wave discharges. Specimens from a sympatric fish community near Manaus (Solimoes). *Left* EOD waveforms; each trace represents 14 ms. *Dashed line* zero potential. *Right* The associated amplitude spectra, with the *ordinate* showing logarithmic amplitude in dB; the *abscissa* logarithmic frequency in kHz. *a Rhabdolichops axillaris; b Sternopygus macrurus; c Eigenmannia macrops; d Distocyclus conirostris; e Eigenmannia virescens.* In all fish the fundamental frequency, or first harmonic, is strongest. Note lacking second harmonic in *d*. Digitized at 12 bit/100 kHz from a Nagra tape-recording (38.1 cm/s). Species names refer to Kramer et al. (1981a)

times their fundamental frequency; they are much less sensitive at still higher harmonics. This was shown by behavioral studies of the threshold intensities for various frequencies (Knudsen 1974), and by jamming avoidance experiments (reviewed in Kramer 1985a). (The tuberous electroreceptors tend to be tuned to the first harmonic, or fundamental frequency, of the EOD; see review by Viancour 1979b).

Correlated with an increasing harmonic content in *Eigenmannia's* EOD is a change of waveform from a nearly sinusoidal oscillation in young fish, to a markedly asymmetric wave characterized by a head-negative baseline with superimposed short head-positive pulses in adult males; all intermediate stages are found. Waveform and harmonic content vary in a sexually dimorphic fashion in adult fish (Figs. 3.15, 3.16).

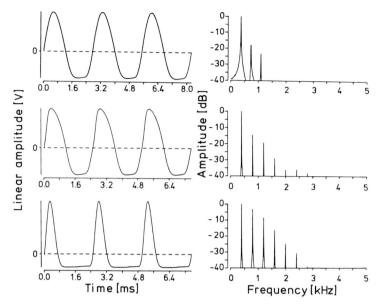

Fig. 3.15. Electric organ discharge of *Eigenmannia*: juvenile, adult female, adult male (from *top* to *bottom*). Waveforms *(left)* and amplitude spectra *(right)*. Note increasing harmonic content and asymmetry of waveform with age, especially in the male sex (Kramer 1985a)

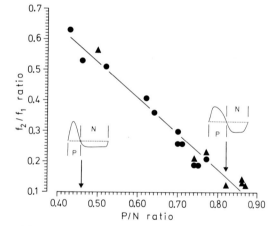

Fig. 3.16. The intensity of the second harmonic, f_2, of *Eigenmannia's* EOD as a function of a temporal waveform parameter, the P/N ratio. The intensity of the second harmonic is expressed as the intensity ratio of the second over the first harmonic, f_2/f_1. The first harmonic, or fundamental frequency, is the strongest component, hence $(f_2/f_1) < 1$. *Circles, E. lineata* [preserved specimens: Zoologische Staatssammlung, München (FRG), ZSM 27156-27160]. Triangles, *E.* sp. 3 (ZSM 27165). Each point is one fish. The same negative correlation holds for both species ($r = -0.983$). The male EOD with its low P/N ratio *(left inset)* is associated with a strong second harmonic, amd a female EOD with its higher P/N ratio *(right inset)* with a weak second harmonic. Young fish all show high P/N ratios; the transition zone to adult male EODs is estimated to be slightly below 0.6, or an f_2 intensity of about 0.4. A least squares regression line is fitted to the data ($y = -1.17x + 1.1$) (Kramer 1985a)

There is no or little DC component in *Eigenmannia*'s EOD, that is, there is no or little net current in the water (review Bennett 1971a; Kramer and Otto 1988). Therefore, the intervals between the zero-crossings of the two half-waves of an EOD cycle, the P(ositive)- and the N(egative)-wave, are almost equal in young fish (hence the P/N-ratio close to, but always below, 1). As the fish grow the P/N-ratio decreases. There is a tight negative correlation between the P/N-ratio and the intensity of the second harmonic ($r = -0.98$). Thus, the simple P/N-measurement allows the quick determination of the intensity of the second harmonic relative to that of the fundamental frequency (Fig. 3.16).

A similar variation in EOD waveform as measured by the P/N-ratio (and most likely also in second harmonic intensity) has first been observed in five *Sternopygus* (Gottschalk 1981). We do not yet know whether this represents a more universal principle in sternopygids.

Eigenmannia's EOD waveform did not change noticeably with gonadal maturity in several females and males (Kramer 1985a). Changes of temperature and conductivity within reasonable limits also barely affect the EOD waveform (Fig. 3.17), although the discharge frequency is strongly correlated with temperature ($Q_{10} = 1.5$; Enger and Szabo 1968).

The independence of *Eigenmannia*'s EOD waveform of conductivity is evidence of there being only one electrically active face in the electrocytes of *Eigenmannia*'s electric organ, that which is innervated by spinal motor nerves. This is in agreement with physiological and anatomical studies (review Bennett 1971a). First-phase voltage which is due to the depolarization of the innervated faces is independent of external resistivity in mormyrid pulse fish as well, but not second-phase voltage generated by the electrically excited opposite faces of the mormyrid electric organ (Bell et al. 1976; Bratton and Kramer 1988; Figs. 3.5–3.7).

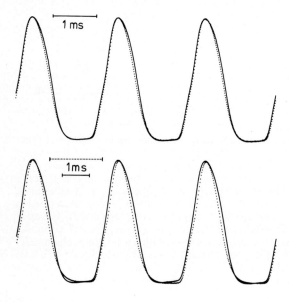

Fig. 3.17. Independence of *Eigenmannia*'s EOD waveform of conductivity (*top*) and temperature (*below*). *Top* Line $= 13\ \mu S \cdot cm^{-1}$; points $= 106\ \mu S \cdot cm^{-1}$ (both at 27° C). *Below* Line $= 19.6°$ C; points $= 32.7°$ C (both at $101\ \mu S \cdot cm^{-1}$). The EOD waves start and end at zero crossings; they are normalized to the same height and to the same number of EOD cycles. *Bars* Time scale (1 ms). On-line digitization at 12 bit/100 kHz (Kramer and Otto 1988)

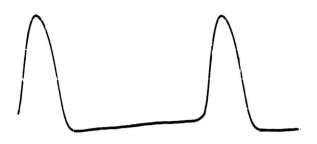

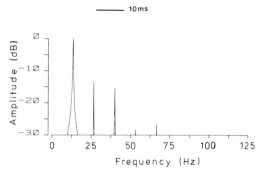

Fig. 3.18. Top EOD waveform of *Distocyclus goachira* (15 Hz-wave species) from the Solimoes, near Manaus. Differential on-line recording on the screen of a storage oscilloscope. Trace starts at zero potential. Head positivity up. Below Amplitude spectrum of digitized waveform

As will be shown in Section 4.2.2.2, female and male *Eigenmannia* show a spontaneous preference for the female EOD when presented with EODs of both sexes, with frequency and intensity cues not being factors; the fish can also be trained to discriminate between these and other waveforms.

The EOD of a single specimen of *Distocyclus goachira* caught near Manaus may be considered intermediate between wave and pulse (Fig. 3.18; this fish was unwounded and had a strong and stable EOD for months). With other sternopygid wave species, it shares the feature of wide, head-positive pulses repeated at a very stable frequency (the latter is known for a few pulse species only; see Sect. 3.2.2.1). With pulse species, however, it shares the features of a very low discharge frequency (15 Hz), and of a pulse duration (20 ms) that is much shorter than the pulse intervals (67 ms). The pulse/interval duration-ratio, or duty cycle, is only 0.3 in *Distocyclus*. (The same figure in the most "male-like" *Eigenmannia* EODs is 0.43). When *Distocyclus'* EOD is played back by loudspeaker the human ear detects its higher harmonics only.

These sternopygid EODs are all basically similar: the fundamental frequency is the strongest spectral component. As the head-positive pulses narrow and the pulse/interval-ratio declines, higher harmonics become stronger (but never stronger than the fundamental frequency). A good example is the maturity- and sex-related change of EOD waveform in *Eigenmannia* (Fig. 3.15). However, in the *Eigenmannia* species complex the EOD cannot be said to be species-specific. As a basis for an intraspecific variability analysis, a taxonomic study of this genus is badly needed (see, for example, Hoedeman 1962a,b for a discussion of geographical populations of *Eigenmannia virescens*, *E. lineata*, and speculations on hybridization with *E. macrops*). The issue

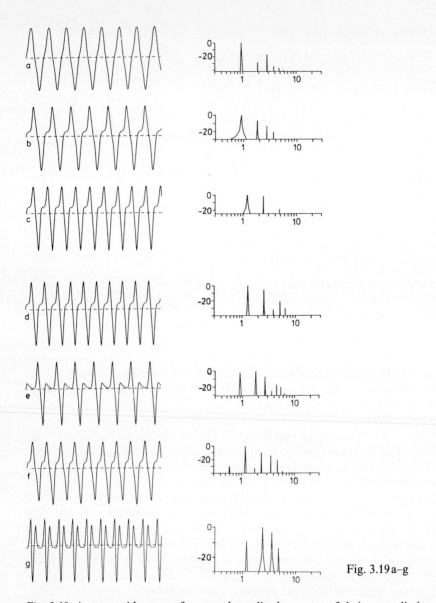

Fig. 3.19. Apteronotidae: waveforms and amplitude spectra of their wave discharges. Specimens from a sympatric fish community near Manaus (Solimoes). *Left* EOD waveforms; each trace represents 8 ms. *Dashed line* zero potential. *Right* The associated amplitude spectra, with the *ordinate* showing logarithmic amplitude in dB; the *abscissa* logarithmic frequency in kHz. *a Apteronotus albifrons*; *b A. hasemani*; *c A. bonaparti*; *d A. anas*; *e A.* sp. 1; *f A.* sp. 2; *g Sternarchogiton natterreri*; *h Sternarchogiton* sp. 1; *i Sternarchorhamphus macrostomus*; *j Sternarchella schotti*; *k Oedemognathus exodon*; *l Porotergus gymnotus*; *m Sternarchorhynchus mormyrus*; *n Sternarchorhynchus curvirostris*; *o Sternarchella* sp. 1. Note that in several species it is one of the higher harmonics that is strongest (in *o*, the fifth). Note also greater variety of waveforms as compared with the Sternopygids (see Fig. 3.14). Digitized at 12 bit/100 kHz from a Nagra tape-recording (38.1 cm/s). Species names refer to Kramer et al. 1981a)

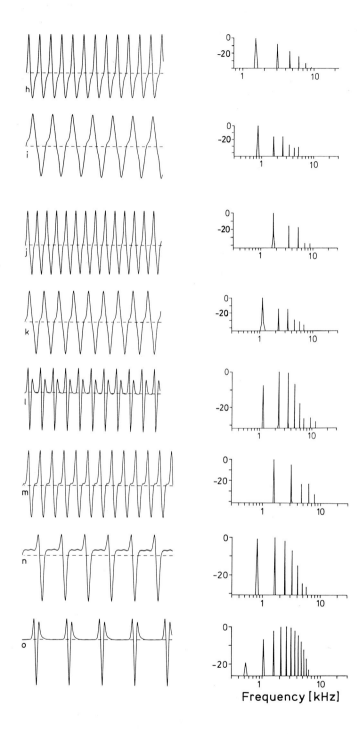

Fig. 3.19 h–o

is made still more complex by Lundberg and Stager's (1985) discovery of considerable microgeographic diversity in *E. macrops* from the Orinoco river.

The electrical "artists" among the wave gymnotiforms, and electric wave fish in general, are the Apteronotidae with their electrocytes being spinal neurons rather than cells of myogenic origin (see review Bennett 1971a). These fish display an unheard of richness in EOD waveforms. As a group, the Apteronotidae discharge at higher frequencies (about 580-1800 Hz; Fig. 3.19) than the Sternopygidae (15 to about 800 Hz).

The stereotypy of EOD waveforms seen in the sternopygids is due to a phase difference of around 0° of higher harmonics (especially the strong second) relative to the EOD fundamental frequency, the first harmonic. Apteronotids, however, may show other phase differences as well, sometimes resulting in characteristic EOD waveforms. For example, in the EOD of the well-known and commercially available black ghost knife-fish (*Apteronotus albifrons*) a phase difference of around 90° is found for the second harmonic. In other apteronotids, such as *Sternarchorhamphus*, the organ discharge consists of head-negative pulses (Bennett 1971a); that is, the fundamental frequency component is of opposite polarity compared with that of sternopygids (a 180° difference in phase). This greater complexity of EODs as found in the apteronotids compared with the sternopygids is due to their entirely different electric organ anatomy and physiology, with considerable differences among species (Bennett 1971a).

One of the evolutionary trends seen in the EOD of Apteronotidae is a tendency to increase the intensity of higher harmonics at the expense of the fundamental frequency component. In one species, *Sternarchella* sp. 1, it is the fifth harmonic which is the strongest (or "dominant", see below) spectral component. In these species, the energy is no longer predominantly concentrated in a single spectral component as seen in Sternopygids (but see the EOD of the *Eigenmannia* male; Fig. 3.15); rather, the energy is distributed over a wide frequency range, similar to pulse species (the -10 dB attenuation level relative to the strongest signal component spans 6 harmonics over almost 3 kHz in *Sternarchella* sp. 1). The salient difference is maintained, however: the signal is harmonic in apteronotids (concentrated in spectral lines), and "noisy" (distributed) in pulse species.

A peculiar EOD is that of *Apteronotus* sp. nov.: all odd harmonics, including the fundamental, are very weak, while even harmonics are strong, with the second being the strongest. Cycle for cycle there is a regular alternation of amplitude (every other cyle being of slightly reduced amplitude; Fig. 3.20).

Three gymnotiform wave species discharging at similar frequencies (560-610 Hz) may differ drastically in harmonic content and waveform (Fig. 3.20). However, even the great differences in harmonic content and dominant frequency components observed in the Apteronotidae do not appear sufficient for a clear species identification in all cases (Fig. 3.21), similar to the situation in the Sternopygidae.

Apteronotids could probably identify the EOD of conspecifics unequivocally from that of other fish with similar frequency if they were able to monitor EOD waveform, for example by the phase differences among a few of

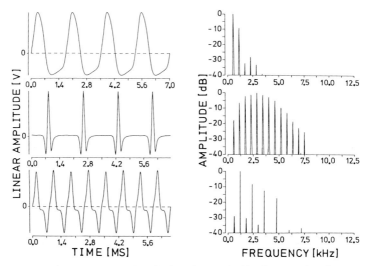

Fig. 3.20. Three gymnotiform wave species discharging at similar frequencies close to 600 Hz; with marked differences in waveform *(left)* and harmonic content, as seen in amplitude spectra *(right)*. Note alternating pattern of low and high intensities in odd and even harmonics, respectively, in bottom example. Top *Eigenmannia macrops*, 561.5 Hz; middle *Sternarchella* sp. 1, 585.9 Hz; below *Apteronotus* sp. nov., 610.4 Hz

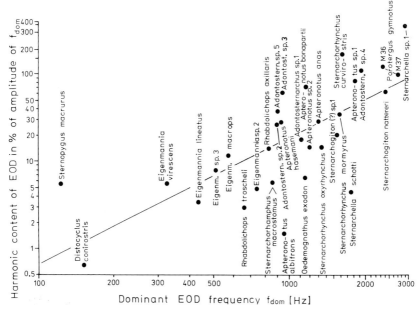

Fig. 3.21. Double log plot of dominant EOD frequency vs harmonic content of EOD for wave fishes. A least squares regression line shows the positive correlation between dominant frequency (or sine wave component of strongest intensity) of an EOD and its harmonic content. In some Apteronotidae, the dominant frequency is a higher harmonic of the fundamental frequency. The *ordinate* shows harmonic content as the summed amplitudes of the other harmonics in percent of the amplitude of the dominant frequency. Each point represents one individual (Kramer et al. 1981a)

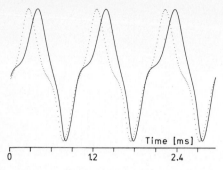

Fig. 3.22. Superimposed EOD waveforms of an *Apteronotus albifrons* recorded at 100 µS · cm^{-1} (26.9° C; *line*) and at 9.9 µS · cm^{-1} (27.3° C; *points*). EODs are normalized to the same height. Head-positivity is up. The animal had time to adapt to a new condition for at least 6 days. Note marked effect of conductivity change on waveform. The experiment was repeated with another fish with similar result. On-line digitization at 12 bit/100 kHz

the strongest harmonics. Whether the fish are capable of doing so is not yet known. The task seems more difficult because EOD waveform changes noticeably with conductivity, at least in *A. albifrons* (Fig. 3.22). It will be fascinating to unravel the mechanism(s) of species identification and reproductive isolation in the Apteronotidae.

3.1.3 Siluriformes

The piscivorous, bottom-dwelling African electric catfish (*Malapterurus electricus* Gmelin 1789) is strong electric. Fish reach 122 cm in length (Lévêque and Paugy 1984), and 350 V discharges (in air) and 356 V (in water) have been recorded (reviewed in Bauer 1968); even 600 V (in air) were reported (Szabo 1970a). The highest value observed by Bauer (1968) was 182 V from a 31.4 cm catfish (water conductivity: 588 µS · cm^{-1}); Rankin and Moller (1986) recorded a maximum voltage of 199 V from a 27 cm catfish (at 180 ± 30 µS · cm^{-1}). These authors described a positive correlation between body length and EOD amplitude, also seen in weakly-electric fishes (for example, Knudsen 1974; Westby and Kirschbaum 1981). The duration of an EOD is 1.3 ms at 28° C and depends strongly on temperature (Bauer 1968). The electric organ is fired by only two giant motor neurons, one on either side of the spinal cord in the first segment of the spinal cord. The electric organ of muscle origin forms a subcutaneous tube enclosing most of the fish (reviewed in Bennett 1971a). Until recently, *M. electricus* was the only catfish known to be electric; two weakly electric catfish species of the genus *Synodontis* from Africa may now be included (M. Hagedorn and T. Finger, unpubl.).

Although the electrocytes are innervated posteriorly, the polarity of the monophasic EOD is almost entirely head-negative in *Malapterurus* (Fig. 3.23). Bennett (1971a) explained this apparent violation of Pacini's rule by a small initial head-positive phase seen only at high gain (Pacini stated that it is

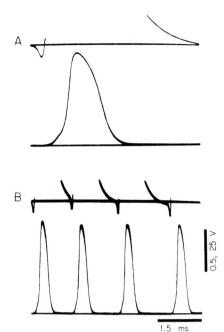

Fig. 3.23. EOD of the electric catfish, *Malapterurus. Baseline* indicates zero volt. Head-negativity upwards. Traces *A,B*, high gain; *upper two traces*, expanded time scale. Note small head-positive potential preceding the main EOD (Bennett 1971a)

the innervated side of a muscle that becomes negative when excited). The physiological mechanism of the catfish's organ thus appears to be similar to other biphasic pulse fish (for example, the gymnotiform *Hypopomus* with a biphasic EOD).

The electric catfish is depicted on numerous murals in ancient Egyptian tombs (for example, around 2750 B.C.; review Zimmermann 1985), sometimes next to the symbol for lightning (Schwassmann 1978b). The first written report about its unusual numbing power, resembling that of the electric ray's that was so vividly described by the ancient Greeks and Romans (Kellaway 1946), is from a Bagdad medical doctor of the 11th century. The Arabic name for the electric catfish is "father of the fire" (Szabo 1970a). Only in the second half of the eighteenth century was the electric nature of the painful shocks delivered by the strong electric fishes which were known at that time discerned in Europe (that is, *Torpedo*, the electric ray; the electric catfish; and the electric eel). These fish played a major role in the development of both the fledgling sciences of electricity and electrophysiology (reviews Kellaway 1946; Szabo 1970a; Wu 1984; Zimmermann 1985).

3.2 Patterns of Spontaneous Discharge Rates

Unlike strong electric fish, weakly electric fish discharge continuously throughout their lives. Wave fish are highly stable in frequency, most pulse fish (except some species among the Gymnotiformes) are not, and show a correlation of discharge rate with overt motor activity (Lissmann 1958). The more intense the motor behavior the higher the discharge rates generally are; lowest discharge rates are shown by resting fish.

Belbenoit et al. (1983) briefly review the evidence for a circadian control depressing global activity (often including the electrical discharge activity) during the rest period, which is the day in mormyriforms, gymnotiforms and the electric catfish. Diurnal *hiding* and *resting postures*, as seen in many species discussed in this book, may be tactics to avoid predators, and may show some characteristics of "sleep".

3.2.1 Mormyriformes

3.2.1.1 Mormyriform Pulse Species (Mormyridae)

Resting Discharge Patterns (Mormyridae). From their field work Moller et al. (1979) concluded that mormyrids tend to be sedentary and solitary during the day and move about during the night. During the day, discharge rates are lower than during the night. This is also observed in large aquaria.

Isolated mormyrids may be completely inactive for prolonged periods of time during the day. The motor activity level observed during the day depends, however, on the species: some species, such as *Brienomyrus niger* or *Pollimyrus isidori*, move little, while others, such as *Gnathonemus petersii* or still more so *Petrocephalus bovei*, tend to be restless. In all species investigated there is a strong correlation of EOD activity with overt behavior. The lowest discharge rates ("basic rates") are observed in immobile fish resting on their fins, or supported by dense filamentous plants if not in contact with the ground, and not moving at all except for their opercula (respiratory movements). Stationary, hovering fish (sometimes also referred to as "resting") may display markedly different EOD patterns, even when inside their shelter (like a tube; see Sect. 4.1.1).

In completely immobile fish, there are usually two or three preferred discharge rates (or modes as seen in histograms of inter-pulse intervals). Interval distributions are always wide. Mean discharge rates are usually below 10 Hz (Fig. 3.24).

The position of modes is shown in Fig. 3.25 in a purely diagrammatic manner in order to facilitate comparison between species. The positions and numbers of modes usually are species-specific and vary only little among individuals of the same species; however, in *G. petersii* a high-frequency mode is present in most but not all individuals (Bauer 1974, Teyssèdre et al. 1987). Between individuals, the lowest frequency mode (long intervals) usually is the most variable as to its numerical value (for example, in *G. petersii* and *P. isidori*). In large species with long discharges the resting EOD activity tends to be of low frequency, with very wide interval distributions and indistinct peaks (for example, *Mormyrus rume* and *Mormyrops deliciosus*), apart from a distinct peak near 100 ms. In *Marcusenius macrolepidotus*, inter-individual

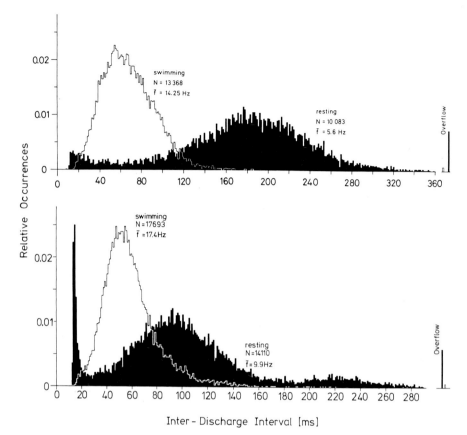

Fig. 3.24. Histograms of inter-discharge intervals of a *Brienomyrus niger* (*top*) and of a *Pollimyrus isidori* (*below*). *N* Total number of EOD intervals analyzed; *f* mean EOD rate (Hz). *Ordinates* give numbers of EOD intervals, normalized to relative occurrences per bin (a fraction of the total, or 1); *abscissas* interval lengths in ms. *Black histograms*, both fish resting (that is, complete immobility except for respiratory movements). Note a prominent mode (peak) at near 100 ms in *P. isidori*, near 180 ms in *B. niger*; both fish also show burst activity (modes at 12-15 ms). *P. isidori*'s resting EOD activity is characterized by a weak third mode of low rate (very long intervals, mode ≈ 220 ms). *White histograms*, both fish swimming. Note the presence of only one mode (peak) in both histograms, contrary to what is observed in resting fish. This mode is between 50-70 ms in *B. niger*, around 50 ms in *P. isidori* (Kramer 1976a, 1978)

variability of resting EOD interval distributions was found to be great (Graff 1986).

As pointed out by Graff (1986), in histograms with two or three peaks standard deviations are proportional to their modes, not only in the *M. macrolepidotus* he studied but also in several other species from the literature. The approximately Gaussian-shaped distribution of a mode is narrow when of high frequency (short intervals) and wide when of low frequency. He found a logarithmic transformation of EOD intervals useful. Instead of bins of constant width (in ms) he used bins growing wider geometrically with interval

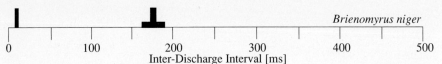

Modes: 14ms, 180 ms (variable); interval range: 10–360 ms; f = 4–8 Hz (Kramer 1976a)

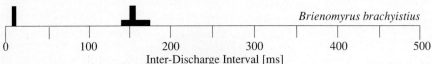

Modes: 11 ms (not present in all fish), 150 ms (+); interval range: 9–600 ms; $f \approx$ 5 Hz (Kramer and Lücker; for fish photograph: see Kramer 1978)

Modes: 12–15 ms, 92 ms, 220–230 ms (variable); interval range: 9–340 ms; $f \approx$ Hz (Kramer 1978)

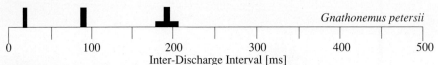

Modes: 22–32 ms (present in most fish), 90–100 ms, (140-) 200 ms; interval range: 18–330 ms (Belbenoit 1972; Bauer 1974; Kramer 1974); $f \approx$ 8–10 Hz

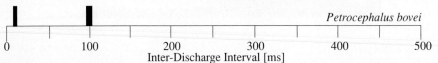

Modes: 12 ms, $\approx$ 100 ms (+); interval range: 10–300 ms; $f \approx$ 10 Hz (Lücker 1982)

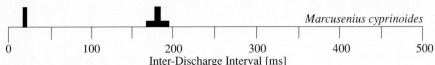

Modes: $\approx$ 20 ms, 160–180 ms; interval range: 13- (at least) 220 ms (Toerring and Belbenoit 1979); f = 8–18 Hz (Toerring and Serrier 1978)

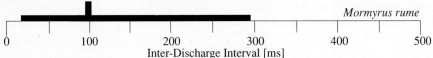

Modes: 105 ms (+), possible further modes indistinct; interval range: 20–390 ms; $f \approx$ 7 Hz (Kramer 1976c)

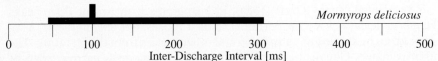

Modes: 90–160 ms (+), possible further modes indistinct; interval range: 50–400 ms; f : unknown (Serrier 1974)

Fig. 3.25

length, affording a good resolution at short intervals, while avoiding the jitter of histogram contours at long intervals.

A resting histogram is also observed in a female *P. isidori*, hiding during the day near a territorial male that will court her the same night. The female's resting EOD activity may even persist during the early period of the dark phase on a spawning night (Bratton and Kramer 1989). The female displays a resting histogram, although the pair clearly detect each other's EODs.

A histogram shows the variation and relative contribution of EOD intervals of different lengths well but sequential information is lost. The EOD sequence displayed by mormyrids has sometimes been characterized as "irregular" or "random", a notion not really supported by the available evidence. In resting *G. petersii*, pairs of adjacent EOD intervals showed a strong tendency of alternating between medium and long intervals in a first-order Markov chain analysis ("joint interval histograms", Bauer 1974; see also Fig. 3.27). This was confirmed by an autocorrelation analysis which also detected the relationships between more distant EOD intervals (see below).

Autocorrelation is a powerful tool to detect periodicities in spontaneous processes, such as neuronal activity, in the presence of blurring noise (for example, Wyman 1965). The mathematical principle involves the multiplication of a time function (see, for example, Box and Jenkins 1976), such as the sequence of EOD intervals (Kramer 1974, 1976a, 1978), or the sequence of EOD rates sampled at equidistant time intervals (Kramer 1979), with itself at a specific delay which is variable. In the case of an EOD interval sequence, the delay is specified as the "number of EOD intervals" (≥ 1; zero delay which yields a correlation coefficient of $+1$, or identity, being of no interest). In a sequence of EOD rates the delay is specified as the "number of sampling intervals" (again ≥ 1, the number being proportional to the time separation, or lag). The integral of the product of the two time functions at each delay converges on zero when there are no periodicities in the time function. If, however, periodicities are present they show up as one or more correlation coefficients significantly different from zero. They are positive ($0 < r \leq +1$) in case of similarity, and negative ($0 > r \geq -1$) in case of dissimilarity relative to the mean at certain delays.

Resting *Brienomyrus niger* display a very simple EOD interval length regulation: runs of six (to sometimes up to 45 in other individuals) intervals are similar in length, with closest similarity between adjacent intervals, and a steady decline of similarity, with an increasing number of intervals separating

---

Fig. 3.25. Purely diagrammatic representation of interval histograms, as shown in Fig. 3.24, of the EOD activity displayed by several mormyrid species during resting behavior, focusing on the peaks (or modes). With increasing interval length the standard deviations of modal means usually become wider (see Fig. 3.24); therefore, short interval modes are represented by a *narrow bar*, medium intervals by a *wider bar*, and long interval modes by *still wider bars*. (+) Mode with very long tail of still longer intervals. Abscissas are only approximate; modes (especially long ones), interval ranges (especially the upper ends), and mean EOD rates (*f*) tend to vary among different individuals, and within an individual with time. 25-27° C

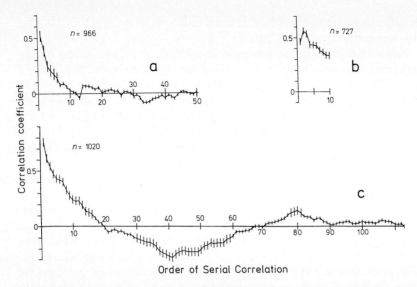

Fig. 3.26. Autocorrelation analyses of the EOD activity of *a* resting *Brienomyrus niger*, *b* resting *Gnathonemus petersii*, and *c* resting *Pollimyrus isidori*. *Ordinates* strength of autocorrelation; *abscissae* lag of autocorrelation, expressed as number of EOD intervals. Correlation coefficients that are statistically significantly different from zero, or no, correlation are marked by *long vertical bar* ($P < 0.01$) (Kramer 1974, 1976a, 1978)

two intervals (Fig. 3.26). Beyond this no periodicity of interval length regulation was detected, the duration difference of two "distant" intervals being, on average, randomly distributed around zero.

P. isidori's EOD rhythm is markedly different from that of *B. niger*. On a run of EOD intervals similar in length (16-21 intervals, with neighboring intervals again being the most similar), there follows a brief period of no relationship to the (arbitrary, moving) first interval. This random fluctuation period is followed by a run of negative correlation coefficients (from 30th-40th to 59th-74th), indicating that a run of long intervals is, on average, "compensated for" by a run of short ones (or, conversely, a run of short intervals, by a run of long ones). In other words, *P. isidori* tends to remain at a certain discharge rate for a relatively well defined number of intervals, and then switches to a higher or lower rate also for a short period of time only, a feature also seen in the "raw data" of a sequential EOD interval plot (Fig. 3.27).

G. petersii is an interesting case, because its resting discharge activity was analyzed by several different methods. Autocorrelation analyses using both the interval criterion for specifying the lag of autocorrelation (as discussed above in *B. niger* and *P. isidori*; Fig. 3.26), and EOD rates sampled at equidistant time intervals (0.2 s; Fig. 4.23) both illustrate that the EOD rate of *G. petersii* remains above or below the mean for a short time (around 2 s) only; after that time EOD rates (and intervals) are randomly distributed with respect to the reference observations. However, EOD interval and rate observations separated by one observation are, on average, more similar than ad-

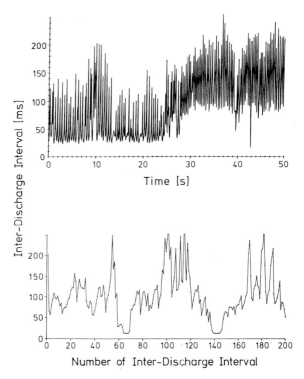

Fig. 3.27. Sequential representations of a resting *Gnathonemus petersii*'s (*top*) and a resting *Pollimyrus isidori*'s (*below*) EOD activity. *Ordinates* give EOD interval length [ms]; *abscissas* time [s] (*top*), or number of interval in the sequence (*below*). In *G. petersii*, note strong tendency of alternating between intervals of about 100 and 200 ms in the second half of the figure, while there are many short bursts (comprising several intervals each) in the first part. In *P. isidori*, note tendency of switching between different EOD rates for bouts of intervals (*top* Lücker 1982; *below* Kramer 1978)

jacent observations. This follows from *G. petersii*'s strong tendency to alternate between "medium" and "long" intervals (see above), a feature not observed in the resting EOD activity of the other species studied. Bauer (1974) and Teyssèdre et al. (1987), who used the technique of "bout interval criterion", suggest that *G. petersii*'s EOD is commanded by two interdependent oscillators, the cycle durations (periods) of which are equivalent to the medium and long EOD intervals. The high frequency burst activity, absent in the resting discharge activity of a minority of individuals, would most parsimoniously be explained not by a third oscillator, but by excitatory input to the higher rate oscillator which normally gives rise to the medium EOD intervals.

This is supported by the observation that short intervals always come in runs (or bouts) during resting activity, and never alternate with medium or long intervals. Also, there is a smooth transition from medium to short intervals.

By a similar line of reasoning, *B. niger* and *P. isidori* both should have only one oscillator governing their EOD rate, despite their complex resting histograms with two or three modes, respectively. EOD rate accelerations and decelerations would be brought about by excitatory or inhibitory input to the oscillator. These inputs appear to be of a cyclic nature themselves, perhaps controlled by a "higher order" oscillator with a much longer period.

A resting *M. rume* displays an interesting variation of short-term compensation mechanism of EOD interval length regulation: adjacent intervals are much more similar than are intervals separated by one (or more) intervals. This results in a rhythm of "short-short-long", or, conversely, "long-long-short" intervals (Kramer 1974).

The interval sequences of all species studied do *not* resemble a random sequence; there is no similarity at all between a resting mormyrid's probability of discharging and, for example, the stochastic series of impulses as detected by a Geiger counter for measuring radioactivity (in spite of, perhaps, a human observer's subjective impression). All species studied show long-term cycles of runs or bouts of intervals (comprising six or more) being, on average, together significantly longer or shorter than the mean; this model may be superimposed by an additional short-term regulation involving close neighbors in an interval sequence (for example, in *G. petersii*).

According to some reports (for example, Bauer 1974; Moller 1980) each individual *G. petersii* has its own characteristic discharge activity during rest, as seen in statistical histograms (like a fingerprint), possibly enabling individual recognition among the members of a group. However, this is difficult to test because any EOD interval histogram is statistically significantly different from a previously recorded one, even when from the same individual at constant conditions, with minimal time separation (Serrier 1982), let alone for differences in the relative height of histogram peaks. Teyssèdre et al. (1987), however, stress the similarity of the resting discharge activity among individual *G. petersii*. The EOD activity also depends on motor activity and time of day (see Sect. 3.2.1.1); mormyrids change their discharge rhythms as soon as they notice another discharging fish (see Sect. 4.2.1), or in response to other supra-threshold stimuli (see Sect. 3.3.1), and sometimes for no apparent reason at all.

As seen in experiments using playbacks of resting EOD activity (Sect. 4.2.1.3), recognition of conspecific EOD patterns is too fast for the fish to build up some equivalent of an interval histogram (which takes about a minute or so for a very crude histogram). Also, the autocorrelation analyses suggest that characteristic interval sequences by which fish might recognize conspecifics are on the order of seconds rather than minutes, because periodicities in the EOD interval sequence tend to die out quickly.

Toerring and Serrier (1978) described a marked temperature dependence of the EOD resting activity of *Marcusenius cyprinoides*; however, it is not specified whether "resting" in their study refers to periods of complete immobility only, excluding, for example, hovering inside a fish's shelter. That this may not have been the case is suggested by Serrier (1982, p. 53). Motor activity depends on temperature.

To study the resting EOD activity of mormyrids may be rewarding for one of several reasons: (1) for the taxonomist (systematist), it may serve as a species character, (2) it indicates motivational state to the ethologist, (3) the physiologist may study the pacemaker mechanisms generating a complex time pattern of EOD intervals, (4) the pharmacologist may use it as baseline data obtained before administering drugs (Kunze and Wetzstein 1988; see Kramer 1984, and unpubl.), and (5) the applied ichthyologist or environmentalist may monitor the quality of tap water by recording the EOD activity of mormyrids (Geller 1984; Kunze 1989).

Motor-Activity Related Discharge Patterns (Mormyridae). Mormyrids are more active during the night than during the day. Daily migrations of mormyrids into and out of a river inlet were observed by Moller et al. (1979) using 12 pairs of recording electrodes distributed in the inlet, the right and left banks of its opening, and in the river. These authors concluded that "during their inactive period during the day, the fish stay inside the inlet, while at night, during their active period, a large number swim out of the inlet to feed along the opening". Several mormyrids hiding individually under protective shelters during daytime departed and returned to their sites at light intensities of around 10 lux (1800 h at dusk, 0623 h at dawn, respectively).

Not only locomotor activity, but also EOD rates observed at night were higher than during day. At their study site, Moller et al. (1979) found a compound mean discharge rate of 15.4 ± 8.2 Hz for individual fish passing by their electrodes during the day (from 0700 through 1600 h), increasing to 26.6 ± 12.7 Hz during the night (1700 through 0600 h). The decrease of EOD rate during dawn was closely correlated with an increase of ambient light intensity (sunrise at 0620 h), while the increase of EOD rate occurred at least 1 h before sunset. Similar effects of changes of light intensity had been observed in the laboratory (Harder et al. 1964; Moller 1970; Bässler et al. 1979).

Isolated, resting mormyrids in the laboratory change their EOD pattern as soon as they begin moving, independent of the time of day (Belbenoit 1972). EOD mean rates observed during locomotor behavior generally are higher compared to the low rates displayed during rest. EOD interval histograms, exhibiting up to three modes during rest in some species, become unimodal and much narrower during swimming. Often the mode observed during swimming does not represent one of the modes observed during rest, but it may coincide with a minimum in the resting histogram (Fig. 3.24).

Accordingly, there is much less interspecific variation in histograms of EOD activity during swimming compared to resting (Fig. 3.28).

Modes are between about 45 and about 100 ms; autocorrelation coefficients are all positive for neighboring intervals, and decline the greater the separation (or lag of autocorrelation; see previous chapter). In *Brienomyrus niger*, such positively correlated runs of intervals ranged from 7-53 intervals, with no interval length regulation beyond (zero correlation, random jitter; Kramer 1976a). In *Pollimyrus isidori*, runs of 16-55 intervals were positively correlated, starting at high correlation coefficients which steadily declined to zero, with an increasing lag of autocorrelation (Kramer 1978).

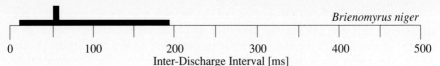

Mode: 50–70 ms; interval range: ≈ 15–190 ms; $f = 10–16$ Hz (Kramer 1976a)

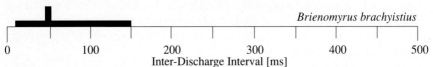

Mode: ≈ 50 ms; interval range: 20- ≈ 150 ms; $f ≈ 19$ Hz (Kramer and Lücker; for fish photograph: see Kramer 1978)

Mode: ≈ 50 ms,; interval range: 12- ≈ 150 ms; $f = 17$ Hz (Kramer 1978)

Mode: ≈ 50 ms; interval range: 26 (20)–75 (120) ms, $f = 16$ Hz (Belbenoit 1972; Kramer 1974, Lücker 1982)

Mode: ≈ 45 ms; interval range: 10–150 ms; $f = 18$ Hz (Lücker 1982)

Mode: 50–70 ms; interval range: 20–170 ms (Toerring and Belbenoit 1979); f: unknown

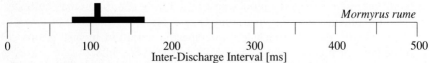

Mode: ≈ 110 ms; interval range: 85–160 ms; $f ≈ 9$ Hz (Kramer 1974)

Mode: 60–80 ms; interval range: 50–270 ms; (Serrier 1974); f : unknown

Fig. 3.28

In two instances, however, this general picture was modified by a superimposed tendency of adjacent intervals to alternate in length (*B. niger*, see Sect. 4.2.1.3; *Mormyrus rume*, Kramer 1974). This is shown by a second coefficient higher than the first one, although both were positive (Fig. 4.40).

Petrocephalus bovei discriminates playbacks of EOD interval patterns of swimming conspecifics from those of two other species (one of them *B. niger*) when also swimming (see Sect. 4.2.1.3). Thus, the fish detect differences among these species' interval patterns, although the differences are seen neither in their histograms, mean EOD rates, nor even autocorrelograms (except that of *B. niger*, see Sect. 4.2.1.3). Autocorrelation which is based on averaging may sometimes fail to detect patterns of species-specific EOD intervals which occur sporadically instead of periodically, or at a great and therefore variable lag (that is, separation). Such characteristic patterns may, however, be obvious from an inspection of interval vs time plots (see Sect. 4.2.1.3).

The EOD activity recorded from *P. isidori* (isolated and paired) differed characteristically for various types of swimming: "slow" and "moderate" swimming, "probing / foraging" and "hovering", when the fish was away from cover like plants and rocks. During these behaviors, EOD mean rates varied from 11 Hz to 28 Hz. "Hiding" (slow fin movement while hiding within a tube or near plants) resembled more closely resting activity, however (Fig. 3.29).

When exploring novel objects in their environment, *Gnathonemus petersii* and *Marcusenius cyprinoides* display various stereotyped 'probing motor acts' close to these objects. During these behaviors fish discharge at a unique and stable, regularized rate, with the EOD interval maintained at 28-30 ms (Toerring and Belbenoit 1979; Toerring and Moller 1984). Sampling the environment at a stable EOD rate is advantageous for precise sensory coding, because of a stable state of sensory adaptation; an increase of EOD rate (as observed here) improves temporal resolution. An EOD rate of about 33 Hz (that is, 30 ms intervals) is still sufficiently below saturation of mormyromasts (Kramer-Feil 1976), the electrolocation receptors (Szabo and Fessard 1974). It is interesting to note that both species (*G. petersii* and *M. cyprinoides*) employ very similar strategies of motor behavior and of discharging their electric organs when showing the behaviors of electrolocating.

Fig. 3.28. Purely diagrammatic representation of interval histograms, as shown in Fig. 3.24, of the EOD activity displayed by several mormyrid species during *swimming* behavior, focusing on the peaks (or modes). Note that in all distributions there is only one mode, contrary to what is found during resting behavior in most species (Figs. 3.24 and 3.25). Interval ranges are indicated by a *horizontal black bar*, disregarding occasional very long intervals. *Abscissas* are only approximate; modes (especially long ones), interval ranges (especially the upper ends), and mean EOD rates (*f*) tend to vary among different individuals, and within an individual with time. 25-27° C

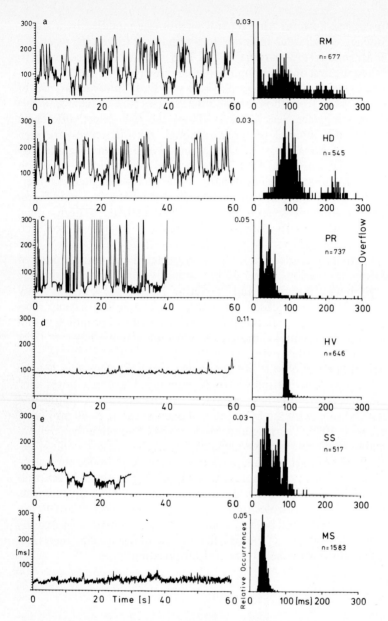

Fig. 3.29. Sequential interval vs time (*left*) and statistical histogram (*right*) representations of *Pollimyrus isidori*'s discharge activity during various type of swimming (c-f) and resting (a,b). *Histograms on the right* represent data to their left. Note variation in EOD activity depending on the type of "resting" or "swimming" (Bratton and Kramer 1989)

3.2.1.2 Mormyriform Wave Species (Gymnarchus)

The continuous, wave-like discharge of *Gymnarchus niloticus* (Fig. 3.8), the only representative of the family Gymnarchidae, ranges from 220-400 Hz at 25-28.5° C (combined data from Lissmann 1958 and Bullock et al. 1975), or 193-326 Hz at temperatures of around 19-21° C (field observations by Moller et al. 1976). *Gymnarchus'* EOD frequency increases with temperature in a regular manner; it decreases with age (Lissmann 1958; see also evidence from field observations by Moller et al. 1976). Its EOD frequency does not change, however, when a resting fish starts moving, and it is unaffected by changes in light intensity, by touch, vibrational stimuli, or the presentation of food. Contrary to the sister group of the Gymnarchidae, the Mormyridae (see previous chapter), *Gymnarchus* studied in the field did not show any EOD frequency increase during the night compared with the day (Moller et al. 1976). Regularity of discharge frequency rivals that of gymnotiform wave fish with the most stable frequencies: the standard deviation of 1000 EOD intervals is only 0.0014% of the mean interval during quiet periods. However, periods of small irregular or more regular frequency modulations (within 1 or 2 Hz) are quite common in *Gymnarchus* (Bullock et al. 1975).

Occasionally, *Gymnarchus* may interrupt its discharge for a brief period for no apparent reason, with no change of frequency (Lissmann 1958). The mean duration of these silent periods was 18.5 s, their frequency 8-9 h^{-1} in one individual while four others did not give any discharge stops during 27 h of continuous observation (Harder and Uhlemann 1967). Bullock et al. (1975) describe a weak residual discharge activity resisting curare, which persists during these pauses; its frequency is near that of the full EOD, its amplitude of only about 1% or less. Usually the full EOD resumes as abruptly as it had stopped. However, preceding onset of the full EOD there may be a transient and intermittent increase of the "miniature EOD" up to 10 or 20% of its amplitude (see also Fig. 1 of Harder and Uhlemann 1967). The resuming main EOD often shows a declining frequency modulation, beginning several percent above and falling within one or a few seconds to the normal frequency for that fish (Bullock et al. 1975).

3.2.2 Gymnotiformes

3.2.2.1 Gymnotiform Pulse Species

Resting Discharge Patterns (Gymnotiform Pulse Species). The EOD waveforms of the four gymnotiform families with pulse discharges are described in Section 3.1.2.1.

Lissmann (1961) observed gymnotiforms in their natural South American habitats. He found that fish were inactive during the day, leaving their hiding places during the night. He was able to observe well-spaced individual *Gymnotus carapo* hiding during the day along the grassy banks of a trench. On successive days fish were found at the same positions. Soon after sunset fish ventured out of their hiding places, for distances only rarely extending beyond 46 to 55 m, mostly moving along the banks.

Steinbach (1970) observed gymnotiforms in the Rio Negro, downstream of its confluence with the Rio Branco. During the day fish (probably *Steatogenys* sp.) aggregated at a rocky area at 16 m depth, which seemed to provide a secure resting area. The fish made migrations of at least 100 m to the surface to a line of submerged trees near the shore (6 m depth), and reaggregated at midstream before dawn. Steinbach speculated that this clear navigational ability might be mediated by feel and smell, with electroreception aiding to an unknown extent.

As in mormyrids, lowest discharge rates are generally observed in resting gymnotiform pulse fish. Unlike mormyrids, intervals are narrowly centered on a mean interval. For example, resting *G. carapo* may display a mean inter-EOD interval of $17.8 \pm SD\ 0.24$ ms over a sample of 1000 intervals; that is, the coefficient of variation is almost as low as 1% in such a sample (Black-Cleworth 1970). The discharge rate is weakly modulated at the low frequency (0.5 to 2 Hz) of respiratory movements in resting *G. carapo*, by 1 Hz or about 2% of the mean EOD rate (Westby 1975c). There are, however, also periods of irregular drifts of discharge rate, the reasons for which are difficult to ascertain (Black-Cleworth 1970; see also Kramer et al. 1981b). Only when fish are excited or start moving do discharge rates increase.

A much narrower variation of EOD rates is observed in a few pulse gymnotiforms, for example *Rhamphichthys*, tentatively identified as *rostratus* (Scheich et al. 1977), *Rhamphichthys* sp. 3, *Steatogenys elegans*, *Hypopomus* sp. 4, *Hypopygus* sp. (Kramer et al. 1981a, b). Unlike *Gymnotus* and most *Hypopomus* species, these fish do not alter their EOD rates in response to stimuli of various kinds (excepting specific, electrical stimuli; see Sect. 4.2.2.1). In *Rhamphichthys* sp. 3, for example, EOD intervals ($n=1024$) varied only by 0.043 ms at a mean of 15.7 ms, or less than 0.2 Hz at 63 Hz (Kramer, unpubl.).

Thus, there exists a sharp divergence among pulse gymnotiforms, with a few species resembling wave species in their extreme regularity of discharge rates, and the others discharging not nearly so regularly. Periodic pulses are a particular harmonic signal type (Scheich et al. 1977). Species that are variable or regular in pulse rate may be found within the same genus, as shown by the genera *Rhamphichthys* and *Hypopomus* (Kramer et al. 1981a, b). EOD rates vary from a few Hz to about 55 Hz in the more variable pulsers, while the extremely regular ones tend to display higher discharge rates (above 55 Hz to about 65 Hz, with the exception of *Hypopomus* sp. 4 with about 11 Hz).

In a sympatric group of pulse gymnotiforms from the whitewater Solimoes river (near Manaus, Amazonas), three ranges of resting EOD rates could be distinguished (Kramer et al. 1981a): (1) the unique electric eel with its extremely low resting rate of weak EODs (which may drop to below 1 Hz), (2) the *Hypopomus* group (about 5-15 Hz) including also *Gymnorhamphichthys hypostomus* of the Rhamphichthyidae (Lissmann and Schwassmann 1965), and (3) more rapidly discharging fish (>30 Hz) like *Gymnotus*, *Rhamphichthys*, *Steatogenys*, and *Hypopygus* (see also Schwassmann 1978b). (The genus *Hypopomus* also includes "class 3" fish; for example,

Schwassmann's 1978a *Hypopomus* sp. r of 66 Hz discharge rate). Because of intraspecific variability, interspecific overlap is so extensive that no single species could be identified on the basis of its resting EOD rate, except the electric eel. Species unknown at present might further enhance the degree of overlap between sympatric species, as well as the tendency of most species to accelerate their EOD rates from time to time, even at rest; and to increase their EOD rates often considerably at night (Schwassmann 1978b).

As shown in Fig. 3.10, by considering two parameters: the frequency of peak amplitude of a species' EOD, as determined from Fourier amplitude spectra of single EODs, and that species' resting EOD repetition rate, the ten species are separated much better than when referring to one parameter alone. A still better result, complete separation, was obtained by Heiligenberg and Bastian (1980) for a community of six pulse fish from the blackwater Rio Negro. In the Solimoes community, which is much richer in species, this method of separation did not, however, appear good enough in most cases; Kramer et al. (1981a) therefore suggested that putative EOD cues reproductively isolating the Solimoes species either are of a still different nature (for example, based on recognizing features of species-specific EOD waveforms; see Sect. 3.1.2.1), or, more likely, rely on a combination of signal properties that may include patterns of EOD rate changes.

Motor-Activity Related Discharge Patterns (Gymnotiform Pulse Species). Lissmann and Schwassmann (1965) and Schwassmann (1971a,b; 1976) observed populations of the sandfish, *Gymnorhamphichthys hypostomus*, in their natural habitats. The sandfish seems well protected from "visually competent predators which abound in South American waters" (Lissmann and Schwassmann 1965), by its habit of burying in sandy banks during the day; it emerges from the sand only 25-40 min after sunset at less than one lux. Fish return to the same locations well before dawn (0320 to 0500 hours).

During the day, when buried in the sand, the fish display a low discharge rate of 10-15 Hz, except for occasional increases. During the two hours prior to actual emergence from the sand the discharge rate of a fish rises to about twice the basic rate. Highest discharge rates are observed during the night: 70-120 Hz are maintained as long as the fish are active and swimming. Intermittent high frequency bursts of more than 200 Hz are observed during feeding or during vigorous swimming movements when the fish are disturbed.

Although weakly electric fish appear to be subjects ideally suited for studies of circadian rhythms, because of their permanent and spontaneous discharge behavior, the sandfish is one of the very few examples which have received appreciable attention in this regard (Lissmann and Schwassmann 1965; Schwassmann 1971a). Schwassmann (1976) writes that "these studies demonstrated that the activity rhythm of the sandfish continues under conditions of constant dim light, with a period depending on the light level, and that the natural day-night cycle entrains the endogenous rhythm to a precisely 24-hour period".

Gymnotus carapo observed in the laboratory also display low discharge rates during the day and higher rates during the night (Dewsbury 1966), similar to a low-frequency *Hypopomus* species with a monophasic EOD (Larimer and MacDonald 1968). Neither fish showed such a strong magnitude of rhythmic effect as did *Gymnorhamphichthys*. In her detailed study of the non-reproductive behavior of *G. carapo*, Black-Cleworth (1970) found a partial explanation for this species' pattern of circadian EOD activity: low discharge rates (35 to 55 Hz) are associated with rest, high discharge rates with swimming behavior (55 to 65 Hz; both behaviors observed at 23 to 26° C). This was confirmed by fish resting at night, and by, very rarely observed, fish swimming during the light phase. Standard deviations of mean EOD intervals were below 0.5 ms, even with samples of 5000 intervals during swimming activity (for example, 16.4 ± 0.39 ms); the coefficient of variation was only 2 to 4%. (This variation represents the whole sample, not adjacent intervals.)

As detailed in Section 3.2.2.1, some pulse gymnotiforms display discharge rhythms, the regularity of which by far exceeds the most regular activity observed in species such as *G. carapo* or most *Hypopomus* species. As in wave fish, and unlike most pulsers, the discharge rates of these fish remain *constant* even when motor activity increases.

Little is known, however, about circadian activity patterns in these ultra-regular fish. *Steatogenys* sp., for example, increased its very regular EOD rate from a mean 36 Hz during the day to a mean 50 Hz at night (Schwassmann 1971a), although Schwassmann (1978a, p. 240) states that "... *Steatogenys* spp. (which) exhibit almost no noticeable day/night differences". This contradiction probably reflects the confused systematics of the Gymnotiformes which is "in dire need of revision; many species have neither been described nor classified" (Schwassmann 1978b). Schwassmann (1984) actually distinguishes two species of *Steatogenys*, *elegans* and *duidae*, although his observation of a frequency increase by 15-20% at night in *S. elegans*, and a more than twofold frequency increase in *S. duidae* do not quite fit either of the earlier observations. Schwassmann's *Hypopomus* sp. r maintained a constant rate (of 66 Hz) day and night (Schwassmann 1978a).

3.2.2.2 Gymnotiform Wave Species

Fish of two gymnotiform families discharge in waves of extremely regular frequency: the Sternopygidae, and the Apteronotidae whose electrocytes are spinal neurons rather than modified muscle cells. EOD frequencies range from 15 to somewhat greater than 800 Hz in the Sternopygidae, and from about 580-1800 Hz in the Apteronotidae Fig. 3.30).

Except for their dependence on temperature, the frequencies of these fish are so constant (Lissmann 1958, Bullock 1970) that they probably represent the most stable neurally paced activities in the whole animal kingdom.

During periods of minimal fluctuation the standard deviation of EOD intervals in *Apteronotus albifrons*, measured between corresponding points of successive EODs, can be as small as 0.14 µs or less over a thousand cycles of a mean of about 1 ms. This is a coefficient of variation (SD/mean $\times$ 100) of only 0.012%; this value was actually limited by the accuracy of the measurement

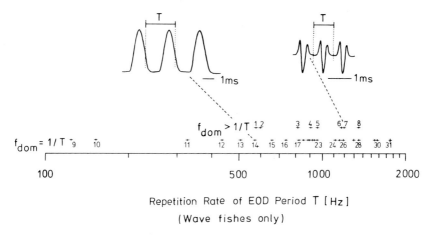

Fig. 3.30. EOD fundamental frequencies, or repetition rates, as observed in a sample of 30 species of wave fishes, sympatric to the Solimoes river near Manaus. *Abscissa* gives logarithmic frequency scale (Hz). *Lower row of numbers* denote species in which the fundamental frequency component was the strongest signal harmonic (*left* waveform example, *Eigenmannia macrops*); *upper row of numbers* species, in which the strongest signal component was a higher harmonic (as in *Sternarchogiton natterreri*, *right* waveform example). *T* One EOD cycle; *numbers* refer to individual species. Each point is one representative individual (Kramer et al. 1981a)

(Bullock 1969). Values almost as low are reported for *Eigenmannia* and *Sternopygus* (Bullock et al. 1975); still lower values were observed in *Eigenmannia lineata*, sampling its EOD over a minute (Kramer 1987, p. 48).

Bullock (1970) compared this extremely high regularity with other examples: classically regular neurons like the crayfish stretch receptor, many primary vestibular afferents, and certain insect optic lobe interneurons have standard deviations of 2 or 3% of the mean interval. This corresponds to "singing the right note within a quarter tone"; *A. albifrons*, however, "sings" its note within a thousandth tone, cycle for cycle over a thousand intervals at least! This variation could be brought about by a temperature fluctuation of only $10^{-3\circ}$ C (at a Q_{10} of 1.7 for that fish); therefore, any long-term estimations (minutes) are extremely delicate (Bullock 1970).

The dependency of the EOD frequency on temperature was studied by many authors, for example, Watanabe and Takeda (1963) and Enger and Szabo (1968) in the wave fish *Eigenmannia*. Most often Q_{10} values of around 1.5 were found (up to 2.0; see review Hagedorn 1986).

Daily migrations have been observed by Steinbach (1970). Data on circadian frequency changes given by Larimer and MacDonald (1968) are inconclusive: the weak frequency fluctuations observed in *Apteronotus (Sternarchus) leptorhynchus* and *Eigenmannia* sp. could be due to minor temperature fluctuations. Schwassmann (1978a) only states that a certain pulse gymnotiform of "steady rate day and night, ... resembles the wave species in this respect"; no publication is referred to.

Frequency modulations in resting, isolated fish do occur but are very rare (*Eigenmannia*; B. Kramer, pers. observ.). There is little or no change of EOD frequency (or of its regularity) associated with motor activity, although an in-depth study seems to be lacking.

In a few wave species a sexual dimorphism in discharge frequency has been observed: *Sternopygus macrurus* (Hopkins 1974b), *Sternopygus dariensis* (Meyer 1983), *A. leptorhynchus* (Hagedorn and Heiligenberg 1985). In *Eigenmannia virescens* a statistical trend for a difference in discharge frequency between males and females was found in large groups of fish (Hopkins 1974a; Westby and Kirschbaum 1981); for small aquarium groups of fish including large males, however, a sexual dimorphism was observed (Hagedorn and Heiligenberg 1985; Kramer 1985a: *E. lineata*). Males display lower EOD frequencies than females in these sternopygids; the reverse holds true for *A. leptorhynchus*, the only apteronotid among these fish. Size is also sexually dimorphic: full grown, adult males are larger than females.

These sex differences in discharge frequency are consistent with experiments using hormone treatment. Androgen steroid hormones decreased and estrogens increased the discharge frequency in *S. dariensis* for several weeks (Meyer 1984; review Dye and Meyer 1986). In *Apteronotus*, however, androgens had little long-term effect, but estrogens decreased the EOD frequency significantly, as would be expected from the observed sexual dimorphism in EOD frequency (M. Leong unpubl., as cited from Dye and Meyer 1986). Additional experiments have shown that these steroid hormones very likely exert their effect directly upon the CNS.

One of the tools used to study the function of neurotransmitters of the CNS are psychoactive drugs which also have powerful effects on mood and behavior. The tranquilizing drug chlorpromazine is known to block the action of a neurotransmitter, the monoamine dopamine (for example, Julien 1988). Dopamine, for which receptors are known only in the brain, plays a key role in motor control (dopamine loss: Parkinson's desease). Dopamine also regulates the secretion of gonadotropic hormone (GTH) from the teleost pituitary: it suppresses the stimulatory effect of gonadotropin-releasing hormone (GnRH) on GTH secretion, and functions as (or is identical with) the gonadotropin release-inhibiting factor (GRIF; review van Oordt 1987). Therefore, dopamine exerts an indirect effect upon the production and secretion of gonadal sex hormones.

The dopamine antagonist chlorpromazine reversibly reduced the EOD frequency by 26.2 ± 9.6 Hz within 3 h of administration in five *A. albifrons* (Kramer 1984). The distribution of monoamines, including dopamine, in the medullary pacemaker and other brain structures of *E. lineata* was studied by Bonn and Kramer (1987).

From the approximate match of "best" frequencies of tuberous electroreceptors with discharge frequencies (Scheich et al. 1973, Hopkins 1976), recognition of conspecifics by their discharge frequency was inferred (Hopkins and Heiligenberg 1978). Only the fundamental frequency component of the discharge would be important, all other harmonics being filtered out by the bandpass properties of the receptors. As becomes increasingly evident, such a mechanism would interfere with the recognition of mates in several species at least, those with sexually dimorphic EOD frequencies. For example, the intraspecific variation of EOD frequencies in *E. virescens* is from about 260-650 Hz at 27° C, or 1.25 octaves.

As also seen in a sympatric community of 30 wave gymnotiforms, a signal's fundamental frequency is of limited use in species recognition because of extensive interspecific overlapping (Fig. 3.30). Similar observations were made by Steinbach (1970) in a Rio Negro community of only ten wave species. Thus, EOD resting frequencies in wave gymnotiforms are species-characteristic but appear to be species-specific only rarely, for example in localities where there are just a few species.

3.2.3 Siluriformes

Laboratory and field observations indicate that the electric catfish, *Malapterurus electricus*, normally firing its powerful electric organ only after detection of a prey fish or when disturbed, may also discharge spontaneously and at low frequency to chase prey ("Scheuchentladungen", Bauer 1968; Belbenoit et al. 1979; see Sect. 3.3.3). Communication signals between electric catfish are nonelectric; EODs have rarely been observed in social encounters (Rankin and Moller 1986; see Sect. 3.3.3).

It is not yet known whether recently discovered weak-electric catfishes from Africa (*Synodontis*; M. Hagedorn and T. Finger, unpubl.) use their EODs in a way more similar to the other weak-electric teleosts: that is, for electrolocation and communication. The electroreceptor periphery is also, as yet, unstudied.

An astounding recent discovery is the tuberous electroreceptor organ in the epidermis of the blind and predatory South American catfish, *Pseudocetopsis* sp. (Cetopsidae), which does not seem to have an electric organ (Andres et al. 1988). This tuberous sense organ bears great morphological similarity to the mormyrid Knollenorgan; therefore, the name *silurid Knollenorgan* has been suggested. The proposed function is the detection of the weak electric fields emanating from the tails of gymnotiforms, which might represent the catfish's main prey. Gymnotiforms with their tails bitten off were found together with this catfish in their natural environment. To date, there is no behavioral nor physiological confirmation of the catfish's proposed sensitivity for high-frequency electric fields.

3.3 Responses to Disturbances (or Food Stimuli)

3.3.1 Mormyriformes

3.3.1.1 Mormyriform Pulse Species (Mormyridae)

The undisturbed EOD activity of mormyrids, as detailed in Section 3.2.1.1, changes in response to several modes of stimulation:

1. Mechanical or vibrational stimuli, such as tapping the aquarium wall, or touching the fish with a glass rod (Lissmann 1958; Bauer 1974);
2. Sound stimuli. A typical response to a sound pulse observed in *Brienomyrus niger* is a transient EOD rate increase from the low and variable resting rate, followed by a discharge stop (Fig. 3.31);
3. Chemical stimuli, such as food odor (Jäger 1974);
4. A sufficiently large, tonic temperature change may have profound effects on the overt motor and the electric behavior of mormyrids (Toerring and Serrier 1978; see also Sect. 3.2.1.1);
5. Likewise, a sufficiently large (tonic) change of light intensity, especially when a threshold of about 10 lx is crossed (reviewed in Moller et al. 1979), and moving visual stimuli (B. Kramer, pers. observ.; for a change in motor behavior: Teyssèdre and Moller 1982). However, an experimental study of the

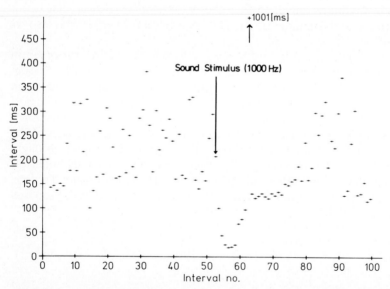

Fig. 3.31. The EOD sound response in *B. niger*. Each point is one EOD. *Ordinate* gives the interval duration; *abscissa* interval number. *Arrow* Onset of a sound pulse of medium high intensity (rise/fall times: 40 ms; plateau time: 400 ms). Note that the first EOD interval after stimulus onset is shortened. A transient EOD rate increase is followed by an EOD stop of 1 s (Kramer et al. 1981b)

effects of light on a possible endogenous, circadian rhythm of discharge rate seems to be lacking. A light flash or a brief switching off of the lights may evoke short-lived responses that may also be conditioned (Mandriota et al. 1965);

6. An impedance change introduced in a mormyrid's electric field, for example, by making or breaking an electrical contact between two conductors in the fish's vicinity, connected by wires outside the water (Fig. 3.32; Szabo and Fessard 1974). The fish feels the change only when it discharges its electric organ. Responses to capacitive impedance changes were observed by Meyer (1982);

7. Artificial, electric stimuli delivered by a pair of electrodes (Harder et al. 1967). Single shocks as well as free-running trains of impulses were both effective in evoking EOD rate changes (Moller 1969, 1970; Serrier 1973, 1974, 1982). Playbacks of a fish's own EODs just to small patches of its skin are also effective stimuli, even at amplitudes on the order of a tenth of a $mV \cdot cm^{-1}$: the EODs were picked up by electrodes, amplified and played back instantaneously, that is, coincident with the fish's EOD (Paul 1972).

Types of response to electrical stimulation were: (1) cessation of discharge, (2) increase of EOD rate, (3) regularization (Moller 1969, 1970; Serrier 1974); and (4) phase-locking behavior (the PLR; see Sect. 4.1.1.1). A regularization is an increase from a low and variable EOD rate to a higher and stable level, with EOD intervals of only one duration (exceedingly small variation). The type of response shown depends on stimulus intensity, duration, and repetition rate.

There may be further modes of stimulation effective in evoking EOD rate changes: for example, magnetic fields. No effect was observed, however, in several *Pollimyrus isidori* despite careful and repeated attempts using several techniques (B. Kramer and U. Heinrich, unpubl.).

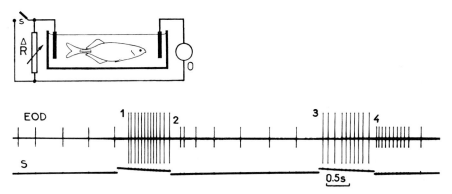

Fig. 3.32. Increase of EOD rate of a *Gnathonemus* in response to the shunting of the field of its electric organ, by an electronic switch (*S, lower trace*). *O* Oscilloscope; *ΔR* variable resistance. As demonstrated by the sudden EOD rate increase of variable latency, it is only when the first spontaneous EOD after closing the switch (at *1* and *3*) occurs that the fish notices the shunt (at *2* and *4* the switch is opened; Szabo and Fessard 1974)

Placing an object in a mormyrid's aquarium (like a stone, a hiding tube, or a metal rod) evoke "probing behaviors", a set of motor patterns correlated with a specific, regular EOD rate described by Toerring and Belbenoit (1979) and Toerring and Moller (1984). It is assumed that these behaviors serve an exploratory purpose, with electroreception being a major sensory modality involved (see Sect. 3.2.1.1).

3.3.1.2 Mormyriform Wave Species (Gymnarchus)

Gymnarchus' EOD is of the constant frequency wave type not easily affected by almost any kind of stimulation (see Sect. 3.2.1.2). A weak electric AC stimulus of approximately fish frequency does, however, evoke the stop response (Szabo and Suckling 1964), which was known already from unstimulated *Gymnarchus*, although at low frequency and not seen in each individual (see Sect. 3.2.1.2). Fish resume discharging in spite of continued stimulation after a fraction of a second or more.

Thresholds for the stop response were lowest for electric stimuli of a frequency range of 50-800 Hz (around 20 $mV_{p-p} \cdot cm^{-1}$; Harder and Uhlemann 1967). These authors found all other kinds of stimulation to be ineffective in evoking a change of EOD frequency, such as: tapping the aquarium walls; vigorously stirring the water; dipping large metal plates into the water; condenser discharges outside the aquarium; moving bar magnets alongside the aquarium. The fish does, however, respond to a magnet or an electrified insulator by a sudden movement (Lissmann 1958; response to magnet confirmed by pers. observ., B. Kramer).

3.3.2 Gymnotiformes

3.3.2.1 Gymnotiform Pulse Species

As explained in Section 3.2.2.1, some pulse gymnotiforms display constant EOD rates which do not change in response to any "ordinary" environmental kind of stimulation, except highly specific electrical stimuli (see Sect. 4.2.2.1). However, pulse gymnotiforms with variable EOD rates, like *Gymnotus* and most *Hypopomus* species, resemble mormyrids: these species apparently respond to all the kinds of stimulation they perceive, and which might alter their state of excitation (Lissmann 1958). These stimuli include mechanical, electrical, visual, light intensity changes, chemical, and auditory (reviewed in Black-Cleworth 1970 for *Gymnotus carapo*).

For example, when the lights were switched off, *G. carapo* which had previously been resting began swimming about its aquarium; correlated with this behavioral change was an increase in its EOD rate from about 40 to 70 Hz (Black-Cleworth 1970). The highest EOD rate observed, a short-lived increase to 200 Hz, was evoked by touching a fish with a glass rod; the presentation of food evoked a maximum frequency of 100 Hz (Lissmann 1958).

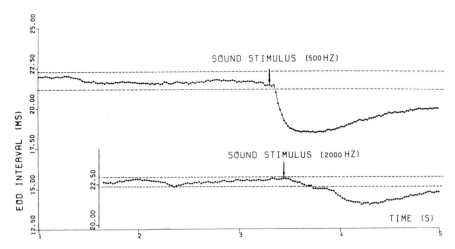

Fig. 3.33. The EOD sound response in *Gymnotus carapo*. Interval duration (*ordinate*) vs time (*abscissa*) plot. Each EOD interval is plotted individually. *Dashed lines* represent the 99%-confidence limits of the EOD interval average before sound stimulation. *Arrow* indicates sound onset. Sound rise/fall times, 40 ms; plateau time, 400 ms; frequency as indicated. *Top* High intensity; *below* medium-high intensity. Note shorter latency and stronger response to stronger sound stimulus (Kramer et al. 1981b)

The magnitude and latency of the "EOD sound response", an EOD rate change to a sound pulse, depends on sound intensity and frequency in *G. carapo* and *Hypopomus* sp. (Fig. 3.33). Except for their higher absolute intensities, threshold-frequency curves of the unconditioned EOD sound response (Kramer et al. 1981b) resemble hearing threshold-frequency curves of other ostariophysine fish, such as the goldfish (see reviews by Popper and Fay 1973, 1984; Popper et al. 1988).

3.3.2.2 Gymnotiform Wave Species

The extreme constancy of discharge frequency of gymnotiform wave fishes of the families Sternopygidae and Apteronotidae is discussed in Section 3.2.2.2. Except for temperature changes no disturbances of any kind affect the discharge frequency of these fishes. Even dying fish which are wounded by netting discharge with a normal frequency for that species. The medullary pacemaker maintains its rhythmical activity at an unchanged frequency, even after surgical removal from the brain in *Apteronotus rostratus* (Meyer 1984).

We do not know, however, whether or not disturbances of various kinds affect the regularity of EOD frequency, if measured at sufficient accuracy (for example, interval for interval at very high resolution, that is, at least 1 MHz).

In spite of their amazing constancy the discharge frequencies of at least some wave species are affected by two kinds of specific stimuli. Weak electrical AC fields of certain frequencies, for example, a frequency close to the fish's own EOD frequency, may evoke the jamming avoidance response, a frequency shift of a few Hz usually increasing the frequency difference (see Sect. 4.2.2.2).

Another type of stimulus affecting *Eigenmannia*'s and *Apteronotus*' EOD frequency is a change of environmental impedance, for example, by short-circuiting two conductors in a fish's electric field (Larimer and MacDonald 1968; their Figs. 9 and 11). Neither the mechanisms of this response nor its functional significance have been elucidated; the implications for the jamming avoidance response are discussed in Kramer (1987, p. 58).

3.3.3 Siluriformes

The piscivorous African electric catfish (*Malapterurus electricus*) discharges its electric organ only occasionally: when hunting prey or when disturbed.

Hunting Prey (Bauer 1968). Mechanical and especially chemical (gustatory) stimuli, as perceived on contact with a prey fish, are the most effective in evoking predatory behavior which is always accompanied by an EOD volley. Taste buds are found all over the body; gustatory stimulation of the tail fin evokes full predatory behavior, as does stimulation of the mouth, or any other region of the catfish's body. Predatory EOD volleys follow detection of the prey and tend to be long (14-562 EODs) compared with those given to disturbances (3-67 EODs; see below). The initial pulse rate is very high (450 Hz at 28° C) and declines rapidly, depending on the size of the prey (which evokes longer volleys if large), the region of the body where the prey first contacted the catfish, and on how successful the catfish is in capturing and swallowing its prey. Catfish with a surgically denervated, inoperative electric organ have a much lower success rate.

The powerful EODs seem to have a numbing effect similar to galvanonarcosis in commercial fishing, when a relatively strong DC current pulsed at a relatively high frequency (30-100 Hz, depending on the size of the fish) is used. (The names of the strong electric rays *Narcine* and *Torpedo* refer to the Greek and Latin words for the stunning or paralyzing effect of strong EOD volleys, respectively).

The discharge activity of *M. electricus* in the wild increases steeply immediately after sunset (Belbenoit et al. 1979), confirming aquarium observations of a predominantly nocturnal activity (Bauer 1968). During the first 6 h after sunset, when electric catfish are especially active, 38% of all EOD volleys recorded in the wild were preceded by a low-frequency pre-volley activity (20-40 Hz). It is assumed that these low-frequency EODs serve a chasing function, evoking startle responses in the inactive, diurnal prey fish (such as cichlids and characids), which could then be detected by mechanical cues (Belbenoit et al. 1979). Similar conclusions based on the observation of occasional isolated EODs occurring without any noticeable stimulus were given by Bauer (1968; "Scheuchentladungen"). Thus EODs would serve an interspecific communication function in *M. electricus*.

In intraspecific aggressive or territorial encounters, EODs are rarely observed; much rarer compared to contacts with prey fish or competitors which do evoke EOD volleys frequently (Rankin and Moller 1986). The intraspecific communication system of electric catfish apparently does not include the use of EODs, except during states of extreme excitation. Such a state may be assumed when one catfish eats a smaller one which continues discharging long after having been swallowed (Bauer 1968). Also, a catfish violently biting another may sometimes give off discharges, although most intraspecific fighting does not escalate beyond ritualized lateral and open-mouth displays without EODs (Rankin and Moller 1986).

Deterrence of Enemies/Competitors. When an electric catfish is prodded with a rod or a brush it may give off a brief EOD volley ("Abwehrsalve"; 3-67 EODs with a mean of 16; Bauer 1968). A catfish squeezed by hand may give off longer volleys (Bauer does not detail the "state of excitation" of the investigator lending his hand). A small catfish which is being eaten by a larger one gives off similarly long EOD volleys (see above).

In their aquarium study on territoriality in electric catfish, Rankin and Moller (1986) also observed EOD volleys from catfish contacting a bichir (*Polypterus palmas*), which was the larger fish. This sympatric carnivore probably is a competitor both for shelter sites and food, and did not seem overly impressed by the catfishes' discharges.

Chapter 4

Communicating with Electric Organ Discharges

During his voyage to the Orinoco river around 1800, Alexander von Humboldt reported communal attack responses in electric eels, and their power to stun prey as large as horses and mules by electric discharges. (He knew about the then recent discoveries concerning the nature of electricity, in which strong electric fish, such as the electric eel, took an instrumental part; Wu 1984; Zimmermann 1985.) Mutual attraction among electric eels when attacking prey and feeding was observed in wading pools (Cox 1938; Bullock 1969).

In weakly electric fishes, Möhres (1957, 1961) observed "tumultuous" EOD rates during the agonistic behavior of the mormyrid, *Gnathonemus petersii*, in aquaria. Through an analogy with the behavior of song birds, he qualified these displays as "electrical territorial singing". Lissmann (1958) largely agreed in stating that *"inter alia*, the electrical discharges have a social significance". He had observed synchronized EOD rate increases and other responses in *Gnathonemus senegalensis*, even when visual and tactile cues were excluded (a pair of fish separated by two parallel, vertically oriented sheets of cloth, 2 cm apart, dividing an aquarium). Clearly, data supporting the fledgling hypothesis of communication by electric organ discharges were necessary.

The first permanent recording of an electric organ discharge interaction in a pair of weakly electric fish is by T. Szabo (in Lissmann 1961). It shows discharge stops of one fish on the discharge accelerations by the other; this observation suggests electrical communication between two species of *Mormyrops* (Fig. 4.1).

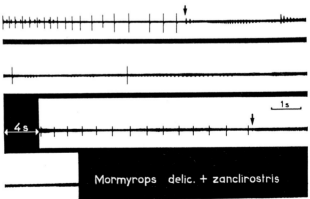

Fig. 4.1. Discharge cessations of a *Mormyrops deliciosus* (*thin spikes*) to high discharge rates of a *Mormyrops zanclirostris* (*arrows*) (Szabo in Lissmann 1961; from Szabo and Moller 1984)

Growing evidence and, finally, conclusive proof of electric communication came only later, with the advent of more sophisticated electronic recording, analyzing, and stimulation equipment. Audio and video recordings on magnetic tape enabled scientists to correlate EOD interactions with overt social behavior. Playback experiments allowed the testing of hypotheses about the "signal value" of certain electrical discharge patterns.

Magnetic tape recorders and laboratory computers were a big and essential improvement over the use of mere monitoring devices without memory, such as amplifier-loudspeaker sets, or conventional oscilloscopes. The first successful studies of electrocommunication (published from about 1970 on) were carried out when such electronic equipment became more affordable or available.

4.1 Electrical and Motor Displays of Communicating Fish

4.1.1 Mormyriformes

4.1.1.1 Mormyriform Pulse Species (Mormyridae)

Non-Reproductive Social Behavior (Mormyridae): Agonistic Behavior. As with most fish, the agonistic behavior of mormyrids can easily be studied in aquaria. A mormyrid isolated in an aquarium for about two days (a "resident") will attack an "intruder", even if it is considerably larger. In aquaria as in nature, fish may compete for similar resources (such as invertebrate food, and hiding places). Therefore, intraspecific aggression is often observed as readily as aggression between members of different species. In every species investigated to date, attack behavior is correlated with a marked acceleration of EOD rate. Maximum EOD rates during aggression are well beyond those observed during isolated swimming, or any other behavior (see Sect. 3.2.1.1).

In *Gnathonemus petersii*, the mean discharge rate recorded during agonistic behavior can reach twice the rate observed in an isolated swimming (about 17 Hz), and three times that of an isolated resting fish (about 8 Hz; Fig. 4.2).

Also, the range of EOD intervals displayed by an attacking fish is much greater than that displayed by an isolated fish. Many of the very long intervals observed during agonistic behavior terminate an attack-associated high discharge rate (see below).

In an agonistic context, an attack will follow when fish (*G. petersii*) display intervals of below 25 ms (Fig. 4.3).

During the course of an attack, the initially low and variable discharge rate accelerates rapidly as the distance from the attacked fish diminishes (Fig. 4.4). The EOD rate associated with physical contact (head butt) varies from 60–80 Hz in most cases (the mode of the distribution is 61 Hz; Fig. 4.5).

The attack-associated acceleration of EOD rate appears to be a ritualized aggression signal (Kramer and Bauer 1976). It could have evolved from an

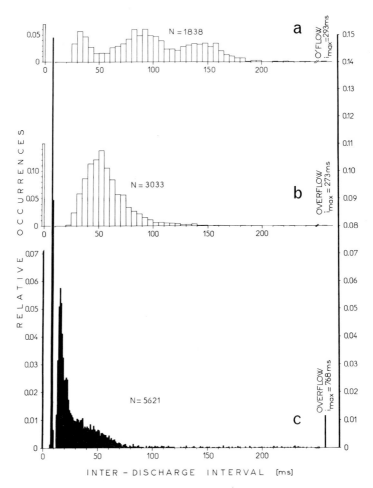

Fig. 4.2. Inter-discharge interval histogram of *a Gnathonemus petersii*: *a* resting in its hiding place; *c* attacking an intruding *Mormyrus rume*; *b* immediately after the intruder had been removed. 3-min recording time in each case. Note the trimodal distribution in *a*, which is bimodal with a gap between the modes in *c*, and the unimodal histogram in *b*. The range of intervals is greatest in *c*; i_{max} longest interval (Kramer and Bauer 1976)

"ordinary" (moderate) EOD acceleration, as observed during the transition from resting to moving (see Sect. 3.2.1.1). While changing its function to an aggression signal this EOD pattern would have become "exaggerated" by ritualization. In its present form the display occurs in a socially significant and well-defined context, the rapid lunge at an opponent in order to butt or bite.

Immediately following a head butt or a bite, *G. petersii* may slide into an antiparallel position relative to the attacked fish, the lateral display. This results in the electric organ of each fish being very close to the other fish's head region, the body part with the highest density of electroreceptors, and highest

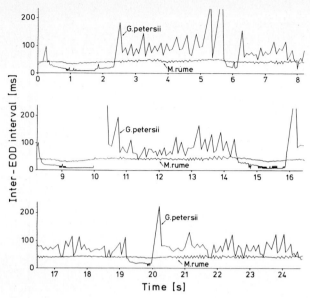

Fig. 4.3. Interval vs time of occurrence plot of the concurrent discharge activities of a *Gnathonemus petersii* and a *Mormyrus rume* during agonistic behavior. The abscissa of each point is the time of occurrence (in s) of each discharge, and the ordinate is the interval (in ms) from the previous discharge (three continuous sections). Note *G. petersii's* high discharge rates, each associated with attack, and their variation in duration and sequence of components which is either a series of equally spaced pulses at 8 or 16 ms, or a paired pulse pattern (Kramer and Bauer 1976)

electrical sensitivity (Harder et al. 1967). During a lateral display, fish emit a high discharge rate of up to 140 Hz, with two types of steady-state activities which may last up to about 4 s: (1) a fairly regular alternation of approximately 16 and 8 ms intervals (paired-pulse pattern); (2) a regular sequence of either 16 or 8 ms intervals. The high discharge rate is terminated abruptly by a discharge break (an interval often longer than 300 ms; Fig. 4.6).

The high discharge rate is remarkable in two ways: (1) a relatively inactive state of motor behavior (the lateral display) is associated with a high intensity electrical display; (2) the two interval levels (8 and 16 ms) may be shown in alternation, or in long sequences of similar intervals of either length. *G. petersii* shows a strong tendency to alternate between two intervals during resting EOD activity as well (intervals of about 100 and 200 ms). As already stated (see Sect. 3.2.1.1), the alternating pattern might be explained by an interdependent activity of two oscillators; if sufficiently excited, they might also generate the alternating pattern during high discharge rates.

Fig. 4.4. A,B. A resident *Gnathonemus petersii* attacking an intruding *Mormyrus rume* with their concurrent EOD activities. *A* Lateral and bottom view, scale marks at 5 cm intervals. *a–q* Intervals between successive pictures = 200 ms (except for *f*, head butt, inserted additionally, and *p* and *q* at 400 ms intervals (see time scale of *B*). *a–b* Approach;

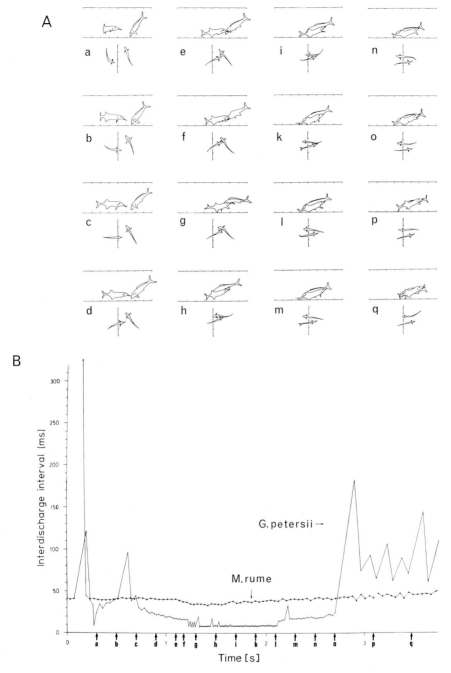

c–f, head butt; subsequently, *G. petersii* glides into an antiparallel position with respect to the *M. rume*. **B** Interval vs time of occurrence plot for the EODs of both fish. Note that *G. petersii's* discharge rate accelerates during overt attack. Physical contact (head butt) occurs just before a stereotyped high discharge rate pattern starts. This pattern is associated with an antiparallel lateral display (Kramer and Bauer 1976)

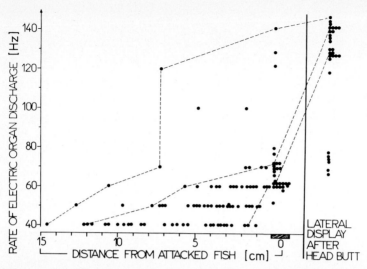

Fig. 4.5. Relationship between an attacking *Gnathonemus petersii's* EOD rate and the distance from its mouth to the body of a *Mormyrus rume*. 28 attacks by 4 *G. petersii*, initial distances ranging from 1–15 cm. The *broken lines* represent three individual attacks. Note that in all but 3 attacks, the EOD rate displayed at zero distance (head butt) is below 80 Hz. During subsequent lateral displays, maximum EOD rates are above 110 Hz in most cases (Kramer and Bauer 1976)

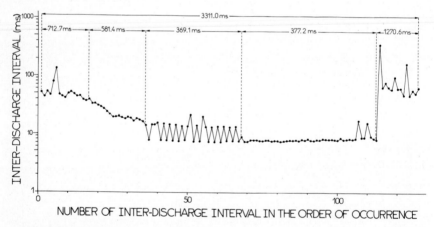

Fig. 4.6. The EOD activity of a *Gnathonemus petersii* attacking a *Mormyrus rume*. Each point is one interval, plotted sequentially on the *abscissa*. The *ordinate* is the length of each interval (logarithmic scale). During an overt attack, interval length rapidly declines. During subsequent antiparallel lateral display, the discharge pattern of a very high rate consists of paired pulses and equally spaced pulses (8 or 16 ms). The high discharge rate display is abruptly terminated by a very long interval; 350 ms in the present case (Kramer and Bauer 1976)

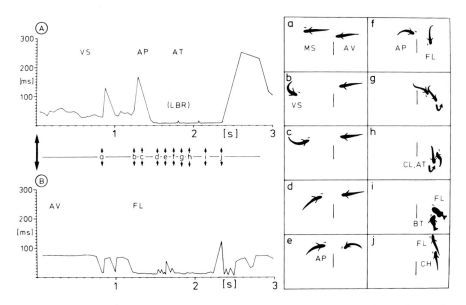

Fig. 4.7. A. EOD activity of a male *Pollimyrus isidori* attacking (*AT*) another male whose concurrent EOD activity is shown in *B*, synchronized with video frames (*a–j*) recorded from underneath during daylight phase. Fish in *A* is to the *left* in *a*, below in *j*. Bar 5 cm reference in tank. *VS* Vigorous swimming; *FL* fleeing; *MS* moderate swimming; *AV* avoidance; *AP* approach; *CL* circling; *BT* immediately before bite; *CH* chase; *LBR* long burst (mean EOD interval: 8.3 ± 1.7 ms, $n = 104$ burst intervals). *Abscissa* time; *ordinate* length of each EOD interval (in ms) (Bratton and Kramer 1989)

Similar high discharge rates during agonistic behavior were observed in intraspecific contests before dominance was established. One fish of a pair was fitted with a fine double strand wire at its tail section for signal separation (Bell et al. 1974). Using a similar technique, Bratton and Kramer (1989) correlated the EOD with video recordings of agonistic behavior in *Pollimyrus isidori*. Synchronized high discharge rates were observed in fish attacking each other, resembling those observed in *G. petersii* (Fig. 4.7).

In contrast to *G. petersii*, *P. isidori* shows its high discharge rate also when resting (intervals of 8–15 ms; maximum species rate). However, these bursts during resting are short (only 7–10 intervals below 30 ms) compared with the high discharge rates observed during agonistic behavior (at least 25 intervals of below 30 ms). Intervals are of similar length in *P. isidori*'s high discharge rates (no alternating pattern). As in *G. petersii*, the motor behaviors of attack occur firmly correlated with EOD accelerations and high discharge rates (Fig. 4.7).

P. isidori's high discharge rates can be maintained for more than 1 s, and may be terminated abruptly by a long interval. During its chase behavior an aggressive fish may show Discharge Breaks (intervals 300–1000 ms) or Discharge Arrests (intervals longer than 1 s), except during its attempts to butt or bite the pursued opponent.

Other species show discharge rate accelerations correlated with overt attacks as well, for example, *Brienomyrus niger* (Kramer 1976a) and *Mormyrus rume* (Kramer 1976c). These fishes' highest discharge rates are, however, considerably lower and of a more transient, not steady-state nature compared with *G. petersii* or *P. isidori*. *M. rume* displays long discharge arrests during agonistic behavior (longer than 7s), even when it is the "resident", larger and dominant fish; the shortest interval during an attack-associated high discharge rate is 34 ms; in *B. niger*, 19 ms.

An aggression signal function of high discharge rates has been tested and confirmed experimentally in *G. petersii* (see Sect. 4.2.1.2). Additional functions, depending on context, are also likely: recognition of sexes; synchronization of mates; behavioral isolation of closely related, sympatric species of mormyrids; spacing out of individuals; defence of territory and nest; displacing other individuals while feeding. Some of these propositions are supported by recent results regarding electric signalling during courtship and spawning in *P. isidori* (Bratton and Kramer 1989; see also below).

Mormyrids may also display EOD patterns signalling appeasement or threat. In classical ethological theory, threat arises from a conflict between attack and escape tendencies when neither can find separate expression; see Manning (1979). The ultimate appeasement signal is, of course, the complete cessation of discharge, as observed in severely attacked and chased fish, sometimes for prolonged periods of time (Kramer 1976b; Bratton and Kramer 1989). A discharge cessation is the perfect "antithesis" (Darwin) of an EOD acceleration; electrically silent fish are attacked significantly less often than discharging fish (Kramer 1976a).

A putative threat signal in *G. petersii* consists of a sudden jump to a stable and relatively high EOD rate of up to 55 Hz, which may be maintained for several seconds, from a much lower and variable EOD rate. This step-like increase of EOD rate with regularization is especially conspicuous when preceded by a Discharge Break (Fig. 4.8). In a chased and fleeing fish, this pattern occurs correlated with each escape reaction to an attacker's repeated attempts to butt or bite (Kramer 1976b).

Attacking *G. petersii* may also receive such a "threat" signal from a conspecific staying put. In these cases, attackers were repeatedly observed to veer away within a few centimeters short of contact with the signaller. Sometimes in fleeing fish there was evidence of a conflict between escape and attacking tendencies struck at a level which seemed closer to the attack tendency; for example, when a regularized, relatively high EOD rate (as typically shown during an escape reaction) was preceded by a rather smooth and rapid, not step-like, decline of interval length, as observed during an attack (Fig. 4.8b, left). A range of conflict behaviors, observed in many vertebrates, is thought to reflect differences in the relative strengths of the underlying tendencies (Manning 1979).

The EOD interval histogram of a chased and fleeing *G. petersii* resembles that of its conspecific attacker by its great range of intervals (including very long intervals), but differs otherwise: it shows only one mode (at the burst activity of the resting histogram, 22–32 ms), while the attacker's histogram

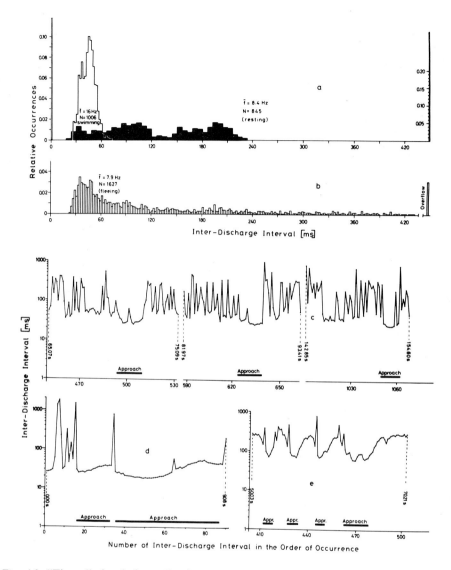

Fig. 4.8. "Threat" signals by a *Gnathonemus petersii* being attacked by a dominant mormyrid. *b, c* EOD activity of a *G. petersii*, being chased and fleeing from an attacking *Mormyrus rume*. *b* Interpulse interval histogram. Note the high variability of interval lengths, resembling the EOD activity of a dominant, attacking *G. petersii* except for the high discharge rates not present here (cf. Fig. 4.2c). The mode of this "fleeing" histogram coincides with the burst interval mode of the resting histogram (Fig. 4.2a). *c* Each point is one interval, plotted sequentially. *Ordinate* length of interval (logarithmic scale). Note that during an approach by the dominant *M. rume*, *G. petersii*'s EOD interval length decreases abruptly to a regularized level (periods without attacks are left out, as indicated). *d* as *c*, but the attacking animal is another *G. petersii* with a "silent" electric organ (rendered inoperative by surgical denervation). Note the step-like interval decreases and long regularizations given by the attacked fish. *e* as *d*, but the approaching, electrically silent *G. petersii* veered away as soon as the attacked fish, which did not move, increased its discharge rate (Kramer 1976b)

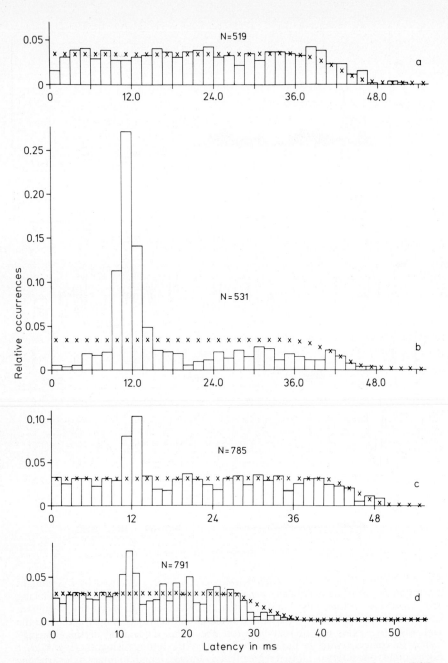

Fig. 4.9. Latency histograms of *Gnathonemus petersii* EODs to *Mormyrus rume* EODs. *a* A latency histogram with both fish isolated does not differ significantly from the expected random distribution *(x's)* of their discharge trains ($P \gg 0.20$; Kolmogorov Smirnov test). *b–d* During agonistic behavior, *G. petersii* exhibits a marked preference to discharge at ca. 12 ms latencies to *M. rume's* EODs. This Preferred Latency Response (PLR) is stronger in weakly and moderately aggressive fish (b, c; $P < 0.01$) than in very aggressive fish (d; $0.05 < P < 0.10$; Kramer 1974)

shows two modes at still higher EOD rates (arising from the 8 and 16 ms intervals during high discharge rate displays). Mean EOD rates are often greater than 30 Hz in attacking fish (highest in the most aggressive fish), and below 10 Hz in fleeing fish.

P. isidori trying to escape from an attacking conspecific display a Medium Uniform Rate of 8–12 Hz, markedly contrasting with the chasing fish's EOD pattern of a much higher mean rate. Except for the context and its shorter duration, this Medium Uniform Rate is very similar to the persistent pattern correlated with courtship and spawning (see below), but normally lasts only for about 20–60 s in an agonistic context and may be followed by a Discharge Arrest (Bratton and Kramer 1989).

Schooling. P. Moller described characteristic spacing patterns for mormyrids which move about in schools, depending in part on the electric sensory modality. The absence of EODs in surgically denervated fish resulted in reduced locomotor activity and the disappearance of certain group behaviors. Electric signals may therefore be considered part of a schooling mechanism that aids the fish in maintaining group cohesion in their turbid environment and during migration at night (Moller 1976; see also Serrier and Moller 1981).

Preferred Latency Responses. One of the most clear-cut examples of communication by electrical signals in mormyrids is the Preferred Latency Response (PLR) which probably represents the most rapid form of communication in the animal kingdom (Fig. 4.9).

The PLR reminds us of entrained signalling, in the form of an alternation of signalling (Popp 1989), such as that which occurs in dueting birds (Thorpe 1972; Wickler 1980; Farabaugh 1982), or frogs calling in synchrony (reviewed in Wells 1977; Zelick and Narins 1985). First described in resident *G. petersii* attacking an intruding *M. rume* (Bauer and Kramer 1974; Kramer 1974), the PLR has also been studied during *intra*specific agonistic behavior in physically restrained *G. petersii* (Russell et al. 1974; the "echo" response). Other species showing the response, which also occurs in the opposite form, the Preferred Latency Avoidance (Kramer 1978), were reported upon later.

G. petersii of both sexes tend to discharge with a latency of approximately 12 ms to the EODs of mormyrids, and to short artificial pulses (phase-locking). The observed latency distribution to the EODs of the mormyrid, *M. rume*, differs significantly from what would be expected if the two discharge trains were independent (Fig. 4.9). *G. petersii* tends to produce preferred latencies in runs of up to 21. Animals which are less aggressive, display a greater number of preferred latencies (Fig. 4.10) and longer runs.

Originally, the significance of the PLR was thought to be avoidance of discharge coincidences (Bauer and Kramer 1974; Bell et al. 1974; Russell et al. 1974; Kramer 1974; see also Sect. 4.2.2.1). The PLR of around 12 ms corresponds precisely to the gap in the EOD interval histogram of a *G. petersii* displaying high discharge rates (approx. 8 and 16 ms intervals; Fig. 4.2c). Bell et al. (1974) give examples of synchronized mutual high discharge rates where 'echoing' seemed present. However, in an agonistic encounter between two *G. petersii* the period of mutual attacking tends to be short because one fish is likely to become dominant very quickly.

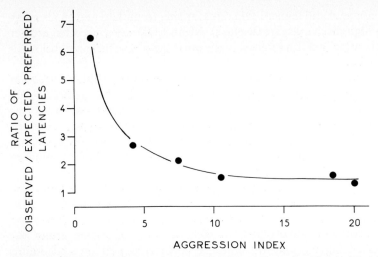

Fig. 4.10. Ratio of observed to expected Preferred Latencies (that is, latencies within 10 to 13.5 ms) of *Gnathonemus petersii* to *Mormyrus rume* EODs, as a function of the aggression index for six resident *G. petersii*. Each point is one fish. The ordinate shows how many times more often than expected, Preferred Latencies occurred. The *aggression index* is the frequency of head butts (each tallied as 1) plus that of approaches (each tallied as 0.5) per min (according to Fiedler 1967; Kramer 1971). Note that the higher the attack rate, the lower the strength of the PLR (Kramer 1974)

The highest proportions of PLRs are shown by fish with the lowest EOD mean rates (that is, fish that attack rarely or not at all). It is in the low-rate EOD activity *between* high discharge rates that by far most PLRs occur (Kramer 1974). This seems reasonable, also in view of Russell's et al. (1974) observation showing that the shorter the latency of a stimulus to an EOD, the less likely it will be echoed (long latencies of fish A not being possible during high discharge rates of fish B). The PLR was also observed in groups of 6 and 14 *G. petersii* without any attack behavior (Serrier 1982).

Russell et al. (1974) and Bell et al. (1974) observed that the reduction of near-coincidences of EODs by the PLR was deceptively low, and Kramer (1976a, 1978) often found even more near-coincidences than expected, in spite of very strong PLR behavior (up to 36% of EODs of freely moving fish within the PLR-range; Kramer 1979).

P. isidori shows latency responses to (1) the EODs of a resting *Brienomyrus brachyistius* (Kramer 1978; probably other mormyrids as well), (2) the EODs of a conspecific (Bratton and Kramer 1989), and (3) artificial stimulus pulses (Kramer 1978; Lücker and Kramer 1981). The response is sexually dimorphic, with males displaying a PLR, females a Preferred Latency Avoidance of around 10–20 ms. Juveniles (up to 6 months, or 6 cm) show neither latency type; their latency distributions do not differ from those expected from two discharge trains with random relationships (Fig. 4.11).

The PLA of *P. isidori* females might perhaps be explained as a strategy for the avoidance of EOD coincidences. This is not possible for the PLR in males, as the burst activity mode of resting histograms and the intervals of the high

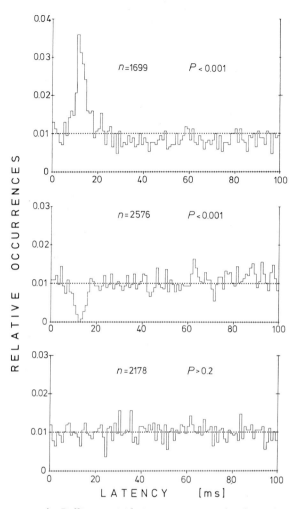

Fig. 4.11. Preferred latency responses in *Pollimyrus isidori* to a constant stimulus pulse rate of 10/s. *A* Preferred Latency Response of a male (full grown at 8.7 cm body length). *B* Preferred Latency Avoidance of the same latency range in a female (8.2 cm). *C* No response in a juvenile (5.2 cm). *Dotted:* expected distributions of random latencies from which observed distributions are significantly different in *A* and *B*, as indicated (Lücker and Kramer 1981)

discharge rate display during aggression are virtually identical with the PLR interval: around 12–14 ms (this is unlike the situation in *G. petersii*). As in *G. petersii*, discharge coincidences are extremely rare for statistical reasons, the very short duration of EODs and their low and variable repetition rates representing a most efficient coincidence avoidance mechanism in itself. The probability of partial or total coincidence of the EOD main phases (50 μs) was determined experimentally in independently discharging pairs of *P. isidori*. Depending on the mean EOD rates, this probability ranged from one to three coincidences in 1000 pairs of pulses (Lücker and Kramer 1981).

An interesting observation is that in several other species a similar form of phase-locking behavior seems to exist, for example, in "baby whales" (genus *Marcusenius*?; Bell et al. 1974). Although less reliable, a preferred latency of around 12 ms was also seen in *B. niger* (Heiligenberg 1976; Kramer 1976a). Serrier (1982) reports a PLR in 11 of the 15 species he tested (not fully documented, except in *G. petersii*); concerning *P. isidori* and *Petrocephalus bovei*, there seems to be some disagreement with the observations of Kramer (1978), Lücker and Kramer (1981), Lücker (1982), and Bratton and Kramer (1989). In *M. rume*, a large species with a long duration pulse, a PLR-like behavior with a latency of around 30 ms sometimes was seen (Kramer 1974). *P. bovei* of both sexes display a clear PLA similar to that of *P. isidori* females (Fig. 4.12).

Russell et al. (1974) studied the physiological properties of *G. petersii's* echo (PLR) response in more detail. The strength of the response depends on stimulus amplitude; the response declines and finally vanishes as the distance between fish is increased to about 30 cm. Head-negative stimuli (or cathodal make) are more effective than head-positive stimuli (or cathodal break). Response latency shifts somewhat with stimulus intensity (shorter latencies

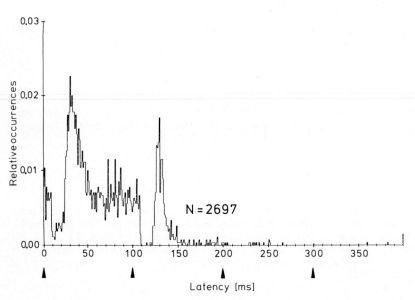

Fig. 4.12. Preferred Latency Avoidance (PLA) in *Petrocephalus bovei* to stimulus pulses of constant rate (10/s), at latencies of 10–20 ms. Males ($n=13$), females ($n=8$) and fish below 60 mm ($n=4$, probably juveniles; smallest fish, 49 mm) all showed the PLA. Stimulation at constant pulse rate ($n=121$ experiments) gave similar results compared with using tape-recorded *P. bovei* EOD patterns for playback ($n=30$). Stimulus pulses were single sine wave pulses (of 143 μs, or 7 kHz), generated by a function generator. *Arrows* Times of stimulus pulses. When fish failed to discharge in response to a specific pulse of the stimulus train (*arrow* at 0 ms on the *abscissa*), an EOD was observed only after the next stimulus pulse (*arrow* at 100 ms), or even later, although this was rare. Note clear PLAs to stimuli at 0 ms (that is, immediate responses) and 100 ms (that is, delayed responses; Lücker 1982)

are obtained at higher intensities). The high response threshold of about 1 mV·cm^{-1} best corresponds to that of mormyromasts with their low sensitivity, while other forms of social EOD responses (for example, stops and regularizations) may be evoked at considerably lower intensities (the Knollenorgane, or large receptors, being 20–30 times more sensitive). Also because of the high efficiency of localized stimulation of *G. petersii's* Schnauzenorgan (the chin appendage, carrying only mormyromast electroreceptors), Russell et al. (1974) concluded that the PLR (echo) is mediated by the mormyromast (medium) receptors.

The most likely explanation of the PLR seems to be a role in sensory gating (although not pulse coincidence avoidance) in the presence of the disrupting sensory input arising from other fishes' discharges. This is suggested by findings concerning adaptation in mormyromast afferences (Kramer-Feil 1976; see Sect. 2.1.3). The PLR / PLA might also have an as yet unknown significance in social communication, and play a role in group cohesion (perhaps the PLR/PLA allows fish to detect and signal "distance to another mormyrid below x cm"; in a pair of *G. petersii* about 25 cm).

Courtship and Spawning (Mormyridae). Field data on the reproduction of mormyrids are exceedingly scarce (reviewed in Kirschbaum 1987). Almost all we know about reproductive behavior comes from aquarium observations of just one species, *Pollimyrus isidori*.

After Birkholz' (1969) chance success in reproducing this species, Kirschbaum (1975; full report 1987) found conditions under which more regular spawnings could be achieved in aquaria. Crawford et al. (1986) found that preceding and during courtship, the territorial male produces an acoustic song. Bratton and Kramer (1989) present the first study of electric signalling during courtship and mating in a mormyrid fish.

Reproductive behavior in the male begins with the defence of a territory and the construction of a nest (Fig. 4.13). Before nest-building, the male may already begin his nocturnal sound production. This song wanes during early courtship and is almost completely absent during the whole spawning period. During the day, the presence of a female is tolerated only when she stays away from the nest region, preferably high up in the water column. There she may coexist with the male for prolonged periods of time, even in rather small aquaria (120 l), especially if given a shelter. An established male shows a PLR, especially during the first few hours after introduction of a female into its tank.

Courtship behavior begins soon after dark (1900 h). Spawning typically starts 2–5 h after dark, continuing for 2–6 h until about 0200 h. During courtship and spawning the female's brief visits (15–25 s) to the male's territory recur every 30–60 s.

Between courtship and spawning bouts, and on nights when the female remains in its hiding region, the male patrols his territory, searching for more nest materials, nudging the nest and foraging for food. He displays a "high sporadic rate sequence" (intervals of 20–80 ms; mean EOD rate about 18 Hz; see Figs. 4.15 and 4.16). Eventually the female advances into the male's ter-

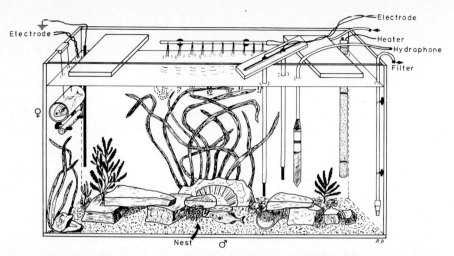

Fig. 4.13. Breeding aquarium (120 l) in which the same pair of *Pollimyrus isidori* have spawned 21 times. The male defends the nest region even against the female which is relatively secure in her hiding tube near the surface. During nocturnal courtship the male directs "courtship attacks" against the female's hiding tube: a rapid lunge at the electrically transparent hiding tube, missing it narrowly, while displaying a high discharge rate (Bratton 1987)

ritory from her surface tube or floating plants, but retreats quickly without contact with the male. Male aggression decreases from 10–20 attacks in the first hour to less than 4 during the next. The male directs "courtship attacks" toward the female hiding in her tube or in surface plants, displaying long high discharge rates, followed by intense singing.

On spawning nights, the female begins a very stable, unique discharge pattern, consisting of nearly equal intervals centered on 100 ms. This occurs in the first few hours of the dark phase, while she is resting in her home region. This EOD pattern of courtship is devoid of short bursts, except during attacks by the male; it continues through the whole courtship and spawning period (see Figs. 4.15 and 4.16). This persistence distinguishes the courtship pattern from the similar hovering and avoidance patterns of non-reproductive behaviors which normally last only 20–60 s, and occur in a chase-flee context not present here.

When the female swims into the male's territory without discharging, the male switches from his "high sporadic rate pattern" to a "medium uniform rate", which resembles the female's courtship pattern (intervals of 80–300 ms; mean EOD rate of 9–11 Hz). When the female swims to a location near the male's bottom hiding place (or into his hiding tube), and waits from 1–3 s, courtship can begin. The two fish circle several times (head-to-tail circling) before the male is able to approach from behind (2–5 s) and position himself laterally to the female (Fig. 4.14). Surprisingly low EOD rates accompany the rather vigorous motor behaviors of head-to-tail circling.

The male then pitches head downward 20° to 40° to the female and quickly rolls 90° to her side while both fish become tightly bound ventrally near the

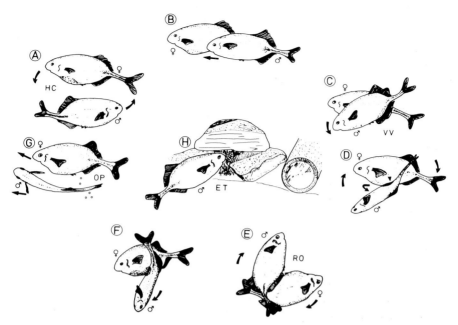

Fig. 4.14. Courtship and spawning in *Pollimyrus isidori* (redrawn from infrared video recordings at night). During courtship (*A–F*) the male approaches the female and *A* head-to-tail circling (*HC*) occurs. *B* The male arrives alongside of the stationary female, *C* becomes coupled vent-to-vent (*VV*), *D* then turns laterally, *E* as both fish pivot around each other, *F* in one complete rotation (*RO*). The male then separates and swims away followed by the female. During spawning the rotation is deleted and the sequence runs from *D* to *G* when the eggs are laid (*OP*). The male then quickly picks up the eggs (*ET*) in his mouth and *H* places them into the nest (Bratton and Kramer 1989)

rostral edge of the anal fin (referred to as "vent-to-vent coupling" by Crawford et al. 1986; and in Fig. 4.14). The female is pushed upward as both move their caudal fins. This causes them to pivot around each other in one full oblique rotation (heads down or tails down both occur) while remaining coupled ventrally. A strong quivering movement of the caudal section accompanies the male's rotation. After completing one rotation (3–4 s), the fish separate and both swim away to their home regions.

While going through the courtship ritual of head-to-tail circling, vent-to-vent coupling, and rotation, the female displays a low to medium sporadic rate with discharge breaks; the male continues his medium uniform rate except for a discharge arrest during vent-to-vent coupling, also observed later when spawning (Fig. 4.15).

On the female's electrically silent (or almost silent) return to her home region, the male switches back to territory patrolling and the associated high sporadic rate.

Spawning immediately follows courtship and usually occurs in the same place. Spawning behavior resembles a short version of courtship behavior: when the male approaches the female from behind, the two fish skip circling, and the male directly advances to position himself obliquely to the female's

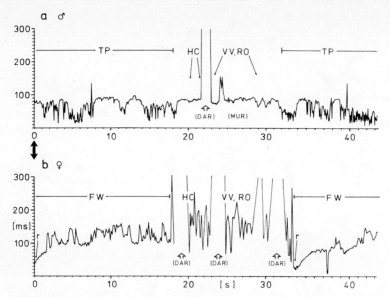

Fig. 4.15. Concurrent EOD intervals during one courtship bout of *a* the male and *b* female *Pollimyrus isidori*. Record begins with the male patrolling his territory (*TP*) while the female waits (*FW*) in her tube (first 18s). The male EOD pattern of a medium uniform rate (*MUR*) continues through head-to-tail circling (*HC*), vent-to-vent coupling (*VV*) and rotation (*RO*), only interrupted by a discharge arrest (*DAR*) during vent-to-vent coupling. On arrival at the courtship site (beginning at 20s) the female displayed a low to medium sporadic rate with many discharge arrests (*DAR*) before returning to her tube (*r* at 34s). *Abscissa* time (in seconds); *ordinate* length of each inter-EOD interval (in ms; Bratton and Kramer 1989)

ventral side. The artistic rotation also observed during courtship is skipped, and the male immediately rolls sideways while forcing the female upward slightly, stimulating her with a caudal fin quiver. This behavior normally lasts 2 to 4 s, during which time the eggs are laid, fertilized, and dropped to the bottom of the tank (1 to 3 s for oviposition; see Fig. 4.14).

The male swims away first, immediately followed by the female returning to her hiding place. The female shows a characteristic caudal quivering motion during the first few centimeters of this return, absent during courtship. The male quickly returns to the spawning site and searches for the eggs, picking up as many as four at a time in his mouth and transferring them to the nest.

The electrical displays of both fish during a spawning bout are very similar to those they showed during the courtship ritual. The male changes his EOD pattern (high to medium sporadic rate) to a medium uniform rate just as the female enters his territory; this pattern continues throughout the spawning behaviors of female spawning site wait, oviposition, and quiver return. The medium uniform rate is only briefly interrupted by a discharge break during vent-to-vent coupling (Fig. 4.16).

The female's EOD activity during spawning also resembles that observed during courtship. On resuming discharging after the female's silent return to

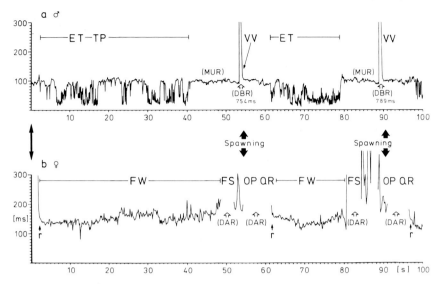

Fig. 4.16. Concurrent EOD intervals during two spawning bouts of *a* the male and *b* female *Pollimyrus isidori*. Record begins with the female just returning to her tube (*r*) and waiting (*FW*), and the male patrolling within his territory (*TP*) and transporting eggs to the nest (*ET*). Note that the male and female spawning patterns closely resemble those observed during courtship (Fig. 4.15). The male shows a discharge break (*DBR*) during vent-to-vent coupling (*VV*); the female arrests her discharge (*DAR*) during part of the spawning sequence. *ET* Egg transport; *FS* female spawning site wait for male; *OP* oviposition; *QR* female quiver return to home region. Other abbreviations and axes as in previous figure (Bratton and Kramer 1989)

her home site, the male immediately switches back from his medium uniform rate to an egg-transport/territory-patrolling activity, with associated high to medium sporadic rate. The female displays her original medium uniform rate while waiting in her territory.

The exact time of onset of the male's medium uniform rate varies with respect to the female's arrival at the spawning site. Part of the time she would arrive after he had begun the medium uniform rate pattern, and spawning would proceed smoothly. When she arrives before the onset of the male's uniform rate, he often attacks her or fails to get into correct vent-to-vent position.

At the completion of spawning, and without disturbances from the male, the female, while remaining in her region, begins displaying a pattern with many rapidly recurring short bursts. These regular alternations of long intervals with bursts were only found to occur in the female. Figure 4.17 presents a summary time table of motor behaviors, and of electrical and acoustic signals during courtship and spawning.

Acoustic signals did not occur during female-male engagement (courting and spawning) except for infrequent courtship attacks. It is not clear why the male started singing again after the end of spawning (after about 0200 a.m. on a spawning night; see Fig. 4.17). One possible explanation is that he would

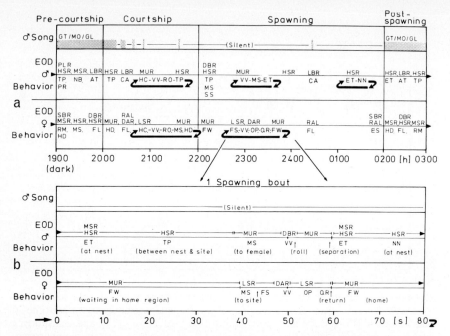

Fig. 4.17. Summary time table showing the relationship of EOD, behavior and song in *Pollimyrus isidori* during *a* the spawning night (lights off at 1900 h) and *b* detail of one spawning bout, as seen in the previous three figures. Motor behaviors: *RM* resting motionless; *MS* moderate swimming; *SS* slow swimming; *FL* fleeing (retreat by vigorous swimming); *HD* hiding (moving fins, near cover); *HC* head-to-tail circling; *FW* female stay (wait) in her territory; *FS* female spawning site wait (1–2 s); *VV* vent-to-vent coupling; *RO* rotation; *OP* oviposition (egg laying); *QR* female quiver return; *ET* egg transport; *TP* territory patrolling; *NB* nest building; *AT* attack; *CA* courtship attack; *ES* end of spawning (female hiding and resting). Electrical displays: *PLR* preferred latency reponse; *HSR* high sporadic rate (20–80 ms intervals); *MSR* medium sporadic rate (80–300 ms intervals); *LSR* low sporadic rate (300–800 ms intervals); *LBR* long burst (>20 intervals of 8–12 ms); *SBR* short burst (<20 intervals of 8–12 ms); *MUR* medium uniform rate (80–300 ms intervals); *DBR* discharge break (0.3–1 s silence); *RAL* regular alternation (of high and low rate); *DAR* discharge arrest (>1 s silence). Song elements: *GT* grunt; *MO* moan; *GL* growl (Bratton and Kramer 1989)

continue to spawn with the same or any other female ready to mate. This would be compatible with a polygynous mating system, let alone for the relatively heavy male investment (nest-building, territorial defense, brood care; see Wilson 1975), and the resources the male holds (territory, nest; resource defense polygyny).

Why should this duality of acoustic and electric communication channels exist? We do not yet know how general acoustic communication is present within the Mormyridae, except for their good hearing (von Frisch 1938; Kramer et al. 1981b; McCormick and Popper 1984) and sound production in *Gnathonemus petersii* (Rigley and Marshall 1973). One reason for the continued (or more recently acquired) capability of acoustic communication, in addition to electric communication, in *P. isidori* may be its weak EOD amplitude. Because of its physiology specialized to produce one of the shortest pulses known from any bioelectric source (17–40 µs for the head-negative main phase; Fig. 3.4) the output of *P. isidori's* electric organ is much weaker than that of *G. petersii* or *Brienomyrus niger* of comparable size (see Kramer 1976b, 1978). The acoustic signal, which might propagate farther than *P. isidori's* weak electrical discharge, would serve well as an "advertisement call" (Crawford et al. 1986). According to Tavolga (1974), "to an aquatic animal, the acoustic sense is probably paramount as a distance receptor".

Gravid females who supposedly do not hold territories could search out males that are established, as indicated by their song. Because the singing activity declines during early courtship and is completely absent during spawning (which may last for several hours), it is the EOD which continues to subserve communication between mates, also during the critical stages of reproductive behavior.

EOD interval patterns probably play an important role in establishing sex identity and acceptance at the time of mating. A specific EOD rhythm (the medium uniform rate) informs the male of the presence of a gravid female, decreases his aggression towards her, and allows her intrusions into his territory while he engages in a similar EOD pattern only when he is ready for courtship and spawning.

4.1.1.2 Mormyriform Wave Species (Gymnarchus)

Lissmann (1958) reviews the biology of *Gymnarchus niloticus*, the only African fish with a constant-frequency wave EOD, including the field observations by Budgett from 1901: *"Gymnarchus* builds a nest of floating vegetation in the middle of which there is a "private pond"; into this are laid eggs, measuring 1 cm in diameter. The parent fish keeps guard near the nest and is, apparently, very aggressive."

Lissmann also noted the "marked cannibalistic tendencies" of *Gymnarchus*, which requires isolation of captive fish. It also seemed to him that two specimens could detect each other's presence at "some considerable distance". Of six pairs of electrodes round the edges of a large, shallow tank, *Gymnarchus* located and attacked whichever pair of electrodes emitted an electric signal (an AC signal of 3 V applied to one electrode pair). Lissmann concludes that "this experiment may indicate the general anti-social tendencies in *Gymnarchus*, but it may be expected that under certain circumstances, e.g., in the breeding season, these tendencies are not indiscriminate. No information is available on this point. A study of electrical and other behaviour responses of a breeding pair may be most rewarding".

Unfortunately, this primitive state of knowledge has not much improved in 30 years. Moller et al. (1976) report that over a period of two weeks, five fish in a permanent pool on an island maintained the same hiding places (during daytime) which they left during night to prey. Fish were recognized by their individual frequencies.

In addition to the spontaneous EOD cessations first described by Lissmann (1958), *Gymnarchus* regularly gives short EOD interruptions when stimulated with an AC current (Szabo and Suckling 1964; Harder and Uhlemann 1967; Bullock et al. 1975), for example, the EOD signal of an isolated conspecific transmitted by wire (Szabo and Suckling 1964). Hopkins (unpubl., in Hopkins 1974c) observed electrical displays related to agonistic behavior. Short EOD interruptions appeared to be correlated with threat behavior, while longer discharge arrests (up to 20 min) were given by subordinate fish when attacked by a dominant one. An EOD arrest reduced the dominant fish's attack rate.

4.1.2 Gymnotiformes

4.1.2.1 Gymnotiform Pulse Species

Non-Reproductive Social Behavior (Gymnotiform Pulse Species): Agonistic Behavior and Electrical Signalling. Gymnotus carapo is a predator, also attacking fish, and lives well-spaced in nature. It is by far the best-studied species (for ecological data, see Lissmann 1961, reviewed in Sect. 3.2.2.1; and Westby 1988).

Black-Cleworth's (1970) investigation into the non-reproductive behavior of *G. carapo* (published as a monograph in *Animal Behaviour*) is remarkable both for the correlation of electrical displays with a wide variety of social (mainly agonistic) behaviors, and for her pioneering and skillful application of modern electronics and computer techniques, which have proven so essential for this field of research.

Black-Cleworth observed *G. carapo* in a large aquarium under seminatural, slightly crowded conditions. Photographs document characteristic postures; a voice description of the behavior was recorded on magnetic tape for correlation with electrical displays. Various forms of lateral and frontal displays were observed, as well as tail curls, jaw locking, roll, serpentining, biting and nipping, nudging and dashes. Further modes of agonistic behavior were described as approach, retreat, withdrawal, and "no response". Westby (1975a) found that the repertoire of behavioral acts of *Hypopomus artedi* resembles that of *G. carapo*.

Black-Cleworth considers the unmodified basic discharge of *G. carapo* (see Sect. 3.2.2.1) an identifying display for signalling species, location, and size of individuals to conspecifics. Sharp increases in frequency followed by decreases to the original level (SIDs), with variable degrees of change and peak frequency, as well as cessations of EOD, occurred during social behaviors, although not exclusively in a social context. Part of the frequency decrease often followed the time course of a decay process (for example, radioactive decay).

The most pronounced SIDs occurred during predation and intraspecific fighting. The greatest degree of change of EOD rate was observed prior to and during actual biting (of both conspecific opponent and prey). Peak frequencies of SIDs can reach 250 Hz. Very similar SIDs are elicited by a variety of stimuli; for example, a fish briefly touched by a glass rod reached a peak frequency of 200 Hz (Lissmann 1958).

The association of sharp EOD rate accelerations with attacks was, however, not absolute (or 100%), as seen in mormyrids (Sect. 4.1.1.1): only 65% of large SIDs displayed by *G. carapo* were associated with attacks, while 63% of all attacks were accompanied by large SIDs. Small SIDs did not show a strong association with any behavioral category. SIDs were shown by biting fish as well as by fish being bitten. The function of this display seemed to depend on context; in addition, SIDs appeared to be graded in intensity, which would further enhance their communicatory potential (but see Westby 1975a; he only saw SIDs occurring in a relatively all-or-none fashion). One function, apparently, is that of threat display, as was also suggested by the observation of dominant fish giving a much higher proportion of SIDs than their subordinate opponents (in agreement with Westby 1975a). Black-Cleworth (1970) regards SIDs as ritualized, emancipated intention displays of biting. (However, for a ritualized behavior, a change in form and emphasis would be required, rendering the original biting-associated EOD rate change more conspicuous by exaggeration, which is not the case in *G. carapo*. Also, the rather loose statistical association of SIDs with attack behavior is not particularly strong evidence for ritualization. Compare, however, this situation with the evidence for ritualization provided by the attack-associated EOD acceleration of mormyrids (Sect. 4.1.1.1)).

In contrast to SIDs, EOD cessations (from 122 ms to 3 min, or more) were common only during agonistic behavior. Cessations often occurred without preceding or following frequency changes; cessations shorter than 1.5 s were termed "breaks", those longer than 1.5 s "arrests". EOD breaks were associated with approach prior to attack. Breaks appeared to be another form of threat display, perhaps indicating less tendency to attack and more tendency to flee than SIDs. Eighty percent of tape-recorded breaks were given by subordinate fish. Westby (1975a) observed an "extreme effectiveness of the *Gymnotus* approach in eliciting an off from its partner".

EOD arrests were associated with "no behavioral act", and retreat; never with attack. All arrests were given by subordinate fish. Discharge arrests appeared to function as appeasement displays. Westby and Box (1970) agree: "turning off" may serve as a submissive signal.

In randomly selected pairs of a total of eight fish, the above-mentioned authors also found that "the frequency of pulse emission measured prior to the experiments [of pairing two isolated fish] was directly related to the extent of aggressive behaviour displayed in social interactions" (for further evidence, see Westby 1975b).

Fighting, so predominant in *G. carapo*, may be restricted to certain times of life or even be largely lacking in other gymnotiform pulse species. Steinbach (1970) observed that *Steatogenys* sp. formed particularly close aggregations, with no apparent intraspecific aggressiveness (confirmed by pers. field observ., B. Kramer).

EOD Synchrony. EOD synchrony may occur in a pair of *G. carapo*, especially when oriented parallel to each other at a separation of more than 10 cm (Valone 1970; Westby 1975a). One fish changes its EOD rate such that it assumes that of its partner. EOD synchrony associated with "orientation behavior" (various stationary positions of a pair of fish, at any distance up to about 25 cm) was also observed in *H. artedi* (Westby 1975a).

Westby (1979) reports that the dominant *G. carapo* of a pair discharging in EOD synchrony has a tendency to place its EOD within the sensitive period of its partner's EOD cycle, whose electrosensory system is probably jammed, while the fish's own system is protected. This interpretation of a preferred latency is supported by the observation of a sensitivity "window" which is phase-locked to the EOD: immediately following an EOD there is a steep behavioral threshold increase to electrical stimuli by about 40 dB (or a factor of 100). This high threshold persists for about 10 ms, then only gradually declining to the point where the next EOD is expected (Westby 1975c; see also Sect. 4.2.2.1).

Jamming Avoidance. When EOD rates are not synchronized, in a pair of fish it is usually the subordinate one which displays the lower frequency (Westby 1979). Because of the properties of the above-mentioned sensitivity window, the lower-frequency fish is in an unfavorable situation at *small* frequency differences: the EODs of the higher-frequency fish slowly scan the EOD cycle from "right to left" (this is in reference to the oscilloscope display with the EODs of the lower-frequency fish triggering the sweep, hence appearing stationary on the screen), or in a "time-negative" direction. From the point of view of the lower-frequency fish, EOD coincidences occur abruptly and without warning because it could not detect the other fish's preceding EODs arriving at successively earlier times during its closed sensitivity gate. Following an EOD coincidence at a small frequency difference, a whole series of EODs will fall into the open sensitivity gate, presumably severely disrupting the lower-frequency fish's electrosensory input. (Gymnotiform fish detect their own EODs by electrosensory feedback via the external medium; they do not seem to possess an internal efference copy, as mormyrids do; see Chap. 2). It is not surprising, therefore, that it was the lower-frequency fish which increased its EOD frequency when the two fish's EODs crossed.

Small frequency differences appear to be less disrupting for the higher-frequency fish, as its EOD cycle is scanned from left to right (in a "time-positive" direction) by its partner's EODs. In that direction, the sensitivity gate opens *gradually*. Westby (1979) describes jamming avoidance manoeuvres and apparent "struggles" for the higher frequency in a pair of *G. carapo* (see also Sect. 4.2.2.1).

Courtship and Spawning (Gymnotiform Pulse Species). To date, neither courtship nor spawning behavior have been observed in any gymnotiform pulse species, except a brief recent description concerning *Hypopomus occidentalis* (Hagedorn 1988; referring also to Heiligenberg unpubl., and Hagedorn unpubl. for *H. brevirostris*). An EOD pattern observed during both aggressive interactions and courtship in *H. occidentalis* is the "Decrement-Burst", a rapid, short-lived EOD-rate increase interspersed during the normal and steady firing of the organ (of around 90 Hz during the night). The duration of a Decrement-Burst is around 13 ms, consisting of a mean 5-6 EODs. These "additional" EODs of very high rate (up to 435 Hz) were always found to be smaller in amplitude than the EODs before and after the display. Courting females exhibit an off-on-off firing pattern: relatively long periods of silence are punctuated by short EOD bursts (Hagedorn 1988).

In dyadic tests for the possession of a single refuge tube in a small aquarium the more dominant males (of a pair) were those of longer EOD duration and higher amplitude compared to their unsuccessful competitors. In these agonistic encounters, the experience of being a winner or a loser affected the EOD amplitude and duration of a male within 2 days (winners changing to more "male-like" EODs; Hagedorn and Zelick 1989).

Hagedorn (1986) gives a useful review of life history data, especially of the sandfish, *Gymnorhamphichthys hypostomus*, studied by Schwassmann (1976, 1978c); and *Hypopomus occidentalis* (see also Hagedorn 1988). Additional data of this nature may be found in Westby (1988).

4.1.2.2 Gymnotiform Wave Species

Non-Reproductive Social Behavior (Gymnotiform Wave Species). Wave fish are remarkable for their extremely stable EOD frequencies, which, apart from temperature, are unaffected by a wide variety of stimuli (see Sect. 3.2.2.2). However, very early in the investigation of the behavior of electric fish, frequency changes during certain real or simulated social contacts were observed in a few species: (1) the jamming avoidance response in *Eigenmannia* (Watanabe and Takeda 1963). (Because of the almost exclusively experimental nature of work on the jamming avoidance response, this topic will be treated in Section 4.2.2.2); (2) During stimulation with weak electric signals, rapid frequency modulations (rises followed by decreases) were observed in *Apteronotus*, called "chirps", "pings", and "spikes" (Erskine et al. 1966; Larimer and MacDonald 1968; Bullock 1969). These displays and their correlation with motor behavior were, however, not studied in detail.

More detailed information is available on *Eigenmannia virescens*. Hopkins (1974a) studied the social behavior in aquaria at night under dim illumination. Isolated fish showed no or little frequency modulations, while socially interacting fish did. Interruptions of the EOD (most often for 20 to 40 ms), often occurring in bouts, were almost exclusively displayed by dominant, attacking fish. Short rises (an increase in discharge frequency followed by a decrease to the resting frequency; duration below 2 s) were also displayed by dominant fish in agonistic encounters, although only rarely. The maximum frequency increase usually was 5 to 10 Hz, but could be as much as 40 Hz. Long rises (from 2 to 40 s) were predominantly recorded from subordinate fish in agonistic encounters, simultaneously with retreat and "no action". The maximum frequency increase was 5 to 20 Hz.

Jamming Avoidance Response. As this behavior has not yet been observed in nature (see above), nor to occur truly spontaneously in aquaria (where it has been studied by using artificial stimulation), it will be dealt with in Section 4.2.2.2.

Active Phase Coupling. The same applies to some fishes' behavior of engaging at certain phase relationships with electric stimuli of suitable frequencies (Sect. 4.2.2.2).

Courtship and Spawning (Gymnotiform Wave Species). Hopkins (1972, 1974b) observed the low-frequency (50–150 Hz) wave fish *Sternopygus macrurus* in its natural habitats, rocky and sandy creeks, during the breeding season. Sexually mature males displayed lower discharge frequencies ($66.8 \pm$ S.D. 13.9 Hz, $n=16$) than females (120.1 ± 19.7 Hz, $n=12$), while juveniles were intermediate (92.3 ± 26.0 Hz, $n=10$; 25°C). Males in breeding condition occupied territories during the day and night; no aggression was observed. The males gave electrical displays (rises and interruptions) to females passing by.

Rises were simple (0.3–2.5 s; 1–43 Hz frequency increase) or complex (0.7–4 s; 1–85 Hz frequency increase). Interruptions (again simple or complex; 0.3–1.7 s) usually occurred within rises, following an increase in frequency and preceding a return to the resting frequency. The discharge did not cease completely during an interruption: low amplitude bursts of pulses usually occurred at low frequency. Rises and interruptions appeared to function solely as courtship, and not agonistic, displays, probably in attracting a female to the male's site, and possibly also in inducing the female to spawn. Also, the female could produce small rises with an average duration of 12.9 s (maximally 12 Hz frequency increase; mostly 7–8 Hz).

Hopkins also observed two isolated male-female pairs which he believed were close to spawning. In both pairs the male's discharge frequency was very close to one octave below that of the female.

Kirschbaum reports breeding in captivity for *S. macrurus* (unpubl.; in Kirschbaum 1987, p. 12; and pers. comm.). Kirschbaum (1983) also achieved reproduction in *Apteronotus leptorhynchus* in the laboratory, but in neither case did he monitor the electrical signals during reproductive behaviors.

Hagedorn and Heiligenberg (1985) observed the courtship and spawning behavior of *Eigenmannia virescens* and *A. leptorhynchus* in aquaria. They found that groups of both species established dominance hierarchies correlated with electric organ discharge frequency, aggressiveness, and size. Several nights before spawning occurred the male modulated its discharge frequency to produce "chirps". Continual bouts of chirping lasted for hours on evenings prior to spawning. In both species the chirp involves a slight increase in frequency followed by a cessation of the dominant frequency (that is, "interruption", see above, in *Eigenmannia*). These electrical signals are thought to play a significant role in courtship and spawning as also inferred from playback.

Kramer and Otto (1988) report spontaneous preference (hence, probably recognition) of the female EOD waveform in *Eigenmannia lineata*, when both male and female EODs were presented simultaneously to isolated fish, by digital synthesis of natural waveforms. Frequency and amplitude cues were ruled out as factors (see Sect. 4.2.2.2).

4.2 Experimental Manipulation of the Electrocommunication System

Communication is, by definition, the complex interaction of two (or more) intact and unrestrained animals; hence, difficult to observe and record, and a nightmare to analyze experimentally.

It is absolutely necessary, then, to reduce the number of variables potentially influencing the behavior of communicating partners as far as possible (without interfering with the investigated behavior), so that observed effects can be attributed to specific causes. Because of the electric fishes' nocturnal habits and a tropical environment hostile to human beings, with waters often being murky, this requirement can usually be only satisfied in the laboratory. Breeding successes with electric fish in a few laboratories have shown that for studying many aspects regarding behavior a good laboratory environment is adequate.

4.2.1 Mormyriformes

4.2.1.1 Ethological Approach

Pairs of Fish Interacting Electrically (Other Cues Being Excluded). The observations of Lissmann (1958) and Szabo (1961) had suggested that two-way communication via the electric channel may occur in mormyrids, even when all other sensory channels are excluded (see beginning of this Chapter). It was only in 1973 that Moller and Bauer gave clear examples of exclusively electrical communication between two intact fish (*Gnathonemus petersii*), in experiments with controlled inter-fish distance.

Two porous ceramic tubes (12×3.7 cm) were horizontally suspended on the two ends of a track, 20 cm above the tank bottom (dimensions: $125 \times 55 \times 60$ cm), by means of a wheeled carriage. One tube was moved back and forth the length of the tank while the other remained stationary at one end. Experiments began only after two fish had accepted the tubes as their permanent day-time hiding places. The tubes were blind-ended at one extremity, with the open endings facing away from each other. This excluded all visual or mechanical interaction between the two fish. The maximal distance between the facing ends of the tubes was 89 cm (later experiments, 4 m; Szabo and Moller 1984).

The EOD rates of both fish were low and variable at an interfish distance of 34 cm; occasional EOD accelerations to about 25 Hz did not affect the other fish's EOD rate (although the fish must have been aware of each other even at the maximal distance of 89 cm; see Squire and Moller 1982). However, when fish A had been moved closer to fish B (to a distance of 24 cm within 6 s), clear EOD rate interactions were observed: "Each time fish B increased its EOD rate to ... 22.5 Hz and more fish A ceased discharging for brief periods. Whenever fish B lowered its EOD rate fish A resumed discharging (...) until fish B again increased its rate which in turn again stopped fish A's EOD temporarily. Fish A resumed discharging when fish B's EOD rate returned to the initial frequency level of ... 7.5 to 10 Hz. On one occasion a simultaneous frequency increase was observed in both fish (...), but fish B's EOD rate exceeded that of fish A. Brief bursts of discharges emitted by fish A prior to and after its temporary discharge cessation (...) were never followed by any significant acceleration or deceleration of fish B's discharge rate. Only fish B's EOD accelerations had the described inhibitory effect on the other fish's EOD activity". (Compare with stimulation experiments using a dipole fish model playing back conspecific pulse patterns, where high stimulus pulse rates evoked similar EOD responses by the resident *G. petersii*; Figs. 4.26, 4.27.) In most cases the EOD increase preceded an EOD cessation (in a few cases this question could not be determined, for technical reasons). An inter-fish distance of 24 cm (where these mutual EOD responses occurred) coincides with the threshold distance for the Preferred Latency or 'echo' response (Russell et al. 1974; see Sect. 4.1.1.1).

These experiments clearly demonstrate electrocommunication in *G. petersii* (Fig. 4.18). However, the observation of EOD rate modulations in a pair of fish and their relationships does not tell us what kind of messages in biological terms was being exchanged, nor can cause always be clearly separated from ef-

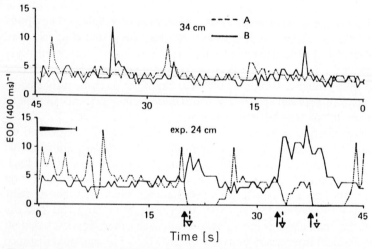

Fig. 4.18. Concurrent EOD rates of two *Gnathonemus petersii* (*A* 15.5 cm, *dashed line*; *B* 9.5 cm, *solid line*) in their daytime shelters, all social cues but electrical being excluded. *Ordinate* number of EODs per successive 400 ms periods; *abscissa* time (s). *Upper diagram*, pre-experiment: at a distance of 34 cm no EOD interaction could be discerned (although later experiments showed that the fish must have been aware of each other; see text). *Lower diagram* After fish A's tube, which was suspended from a wheeled carriage, had been moved closer to fish B (to a distance of 24 cm during 6 s; *horizontal wedge-shaped symbol*), clear EOD rate interactions occurred. The most frequently observed EOD interaction was of the "frequency increase – frequency decrease" type (*vertical pairs of arrows*). Later experiments (Russell et al. 1974) showed that at an inter-fish distance of 24 cm, the threshold of the Preferred Latency Response had been crossed (Moller and Bauer 1973; from Szabo and Moller 1984)

fect in a two-way runaway process of communication. Therefore, other types of experiments were also necessary: for example, the study of the effect of electric signals on motor behavior, and denervation and stimulation experiments ("open loop" experiments).

Denervation Experiments. Moller (1976; see Sect. 4.1.1.1) and Kramer (1976a) have studied the role of electrical signals by using "electrically silent" mormyrids. As shown by Belbenoit (1970), the electric organ of mormyrids can easily be silenced by surgical denervation (lesion of the spinal cord just cranial to the caudal peduncle). Successfully operated fish swim normally, because strong, specialized tendons traversing the electric organ connect the tail fin to powerful trunk muscles (see, for example, Bruns 1971). Like other fast-swimming fish (for example, pike or tuna), mormyrids get most of their thrust by lateral strokes of the big, often sickle-shaped tail fin (see also Blake 1983a).

If EODs play an important role in communication, the social behavior of an intact mormyrid towards an operated, electrically silent partner should be affected by the partner's lack of electric organ discharges. In order to recognize an effect, the behavior displayed towards an operated conspecific was compared to that shown towards an intact conspecific. A heterospecific *Brienomyrus niger* was used as a third type of stimulus fish, for information

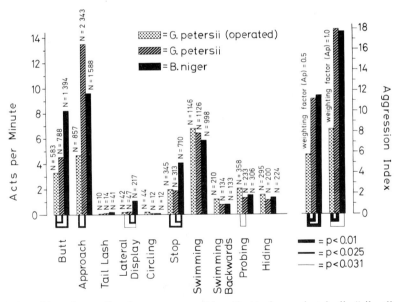

Fig. 4.19. A resident, intact *Gnathonemus petersii* ($n=7$) attacks an electrically "silent" conspecific (with a denervated electric organ) significantly less often than an intact conspecific, or a mormyrid with similar discharge, *Brienomyrus niger*. The *aggression index* combines the scores for *butts* (weight = 1; total $n=2765$) and *approaches* without contact (total $n=4788$). Note that whatever weight is given to approaches (0.5 or 1) barely affects the result: the electrically silent conspecific is attacked least while there is no significant difference between the attack frequencies for the two electrically intact fish (Kramer 1976a)

about the species-specificity of the fishes' electrical activities might also be obtained by observing the experimental subjects' responses.

The experimental subjects (*G. petersii*) displayed significantly fewer butts toward an electrically silent conspecific than they did in the presence of an intact conspecific, or an (intact) *B. niger* (Fig. 4.19). (Each of these stimulus fish was used only once a day, in random sequence, and presented to the experimental subject in its home tank for 5 min. The interval between the three experiments in a day was 4 h).

There was no significant difference between the results for the intact *G. petersii* and *B. niger* stimulus fish. The approach score was also lowest in the presence of the silent *G. petersii*, being significantly different from that observed during an intact conspecific's presence.

In order to simplify the results concerning aggression, an "aggression index" was calculated by giving an approach half the weight of a butt (Fiedler 1967). Also other weighting factors for approach were used: 0 and 1 (Fig. 4.19). A significantly lower rate of attacks on the electrically silent conspecific, as compared to the rates for the two electrically intact types of stimulus fish, was found independent of the weighting factor.

The effect of the stimulus fish may also be expressed in terms of the mean rate of all components observed. An isolated *G. petersii* displayed a very low activity of approximately 0.5 acts/min (hiding, probing, swimming, swimming backwards, stop). This rose to a mean 22.2 acts/min in the presence of a silent conspecific. Both an intact conspecific (29.8 acts/min, $P<0.01$) and a *B. niger*

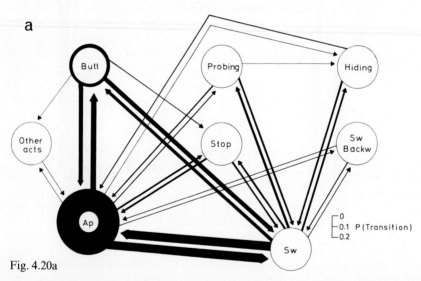

Fig. 4.20a

Fig. 4.20. Transition diagrams for agonistic behavior sequences, displayed by *Gnathonemus petersii* towards one of three types of stimulus fish. *a* Behavior displayed towards an intact conspecific, *b* towards a silent conspecific, *c* towards a *Brienomyrus niger*. *Width of arrow* is proportional to transition probability as shown. *Ap* Approach; *Sw* swimming; *SwBw* swimming backwards; *Other acts* scores for tail lash, lateral display and circling combined (Kramer 1976a)

(32.1 acts/min, $P<0.025$) evoked significantly higher mean rates of components as compared to the silent conspecific. The difference between the two electrically intact types of stimulus fish was not significant.

The effect of communication may be seen by a change in behavior. The analysis of component frequencies (Fig. 4.19) is often too crude a method to describe fully or even to detect a change. By studying transition frequencies from one behavior to the next, differences in the facilitated combinations of motor patterns may be revealed (Baerends et al. 1955), even when the component frequencies displayed by an animal under different conditions are identical.

In the transition diagram, which shows the behavior of *G. petersii* towards an intact conspecific, the predominant importance of the transitions from and

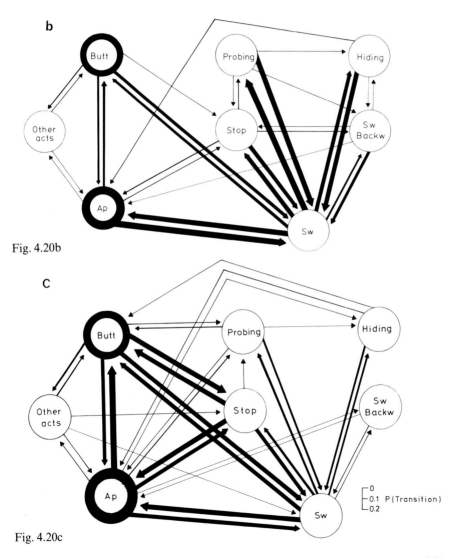

Fig. 4.20b

Fig. 4.20c

to, as well as within, social behaviors is immediately visible (Fig. 4.20). Butt and especially approach activities were extensively performed on their own. The importance of transitions between non-social behaviors is much greater during a silent *G. petersii*'s presence. Superficially, the transition diagram for the *B. niger* experiment resembles that for the experiment with an intact conspecific by the small frequencies of transitions between non-social behaviors. A conspicuous difference between the *B. niger* experiment on one hand, and the experiments with both types of conspecifics on the other, is the importance of the Stop node, which competes with the Swimming node in its function as the main link between social and non-social behaviors.

In neither case can the behavior displayed by *G. petersii* be described by a random model: that is, the probability of one act depends on (at least) the preceding act (difference from random model significant at $P<0.001$, χ^2-test). But are the three distributions different from each other? Only differential effects of the three types of stimulus fish on the behavior of the experimental *G. petersii* would be convincing evidence of communication by electrical signals.

This question was addressed by comparing the distribution of following acts on a given act, found in one type of experiment, with the same distribution from one of the two other types of experiments, in as many $2 \times k$ contingency tables as were needed for comparing all combinations of $k=8$ acts and three stimulus types (for method, see, for example, Siegel 1956). For every behavior and each combination of experiments, a significant difference emerged: that is, the three distributions of transition frequencies are not only different from those expected from a random model, but also fundamentally different from each other (see Kramer 1976a, Table 4).

Statistical information theory allows one to find out more about the differences between the transition frequencies the experimental subjects showed in the presence of the three types of stimulus fish (good introductions into the theory are found in Shannon and Weaver 1949; Quastler 1958; Attneave 1974; Wilson 1975). In particular, a measure of the reduction of uncertainty which is associated with the occurrence of a behavior in a behavioral sequence may be determined, as well as the amount of internal constraint in the system which causes the sequential dependencies, shown to exist by the inadequacy of a random model to explain the observed transition frequencies.

From the component frequencies the average information H (or uncertainty, or entropy) in bit which is associated with the occurrence of a behavior x may be estimated according to the Shannon-Wiener formula:

$$H(x) = -\sum_i p(i)\log_2 p(i),$$

where $p(i)$ is the probability of each behavior x_i. Table 4.1 shows the results obtained for the three types of experiments. By comparing these values with the theoretical maximum which is $\log_2 n$ (n equiprobable categories of x), the relative entropy R is obtained:

$$R(x) = H(x)/\log_2 n.$$

Table 4.1. Information theory analysis of the behavior displayed by *Gnathonemus petersii* in the presence of one of three types of stimulus fish (*cs* conspecific, or *Brienomyrus niger*) (Kramer 1976a)

	Experiment with		
	Intact cs	Electrically silent cs	B.niger
Uncertainty of a behavior $H(x)$ in bit	2.25	2.68	2.57
Relative uncertainty of a behavior $R(x)$	0.75	0.89	0.86
Redundancy $C(x)$	0.25	0.11	0.14
Joint uncertainty $H(x, y)$ in bit	4.21	4.69	4.89
Information shared $T(x; y)$ in bit	0.28	0.65	0.24
Conditional uncertainty $H_x(y)$ in bit (ambiguity)	1.96	2.01	2.32
Coefficient of constraint $D(x; y)$	0.12	0.24	0.09
Relative uncertainty of a behavior $R_x(y)$ (considering first order transitions)	0.65	0.67	0.77
Redundancy $C_x(y)$ (considering first order transitions)	0.35	0.33	0.23

Table 4.1 shows that the uncertainty associated with the occurrence of a behavior was considerably smaller in the experiment with an intact conspecific [$R(x)=0.75$, or 75% of the maximum uncertainty] as compared with the two other experiments, where the relative entropies are rather similar (0.89 and 0.86). In other words, the redundancy C (which is defined as $C=1-R$) of the three behavior sequences was 25%, 11%, and 14%, respectively.

These estimates of the average information $H(x)$ per component or of the relative entropy $R(x)$ are overestimates, since a random model could not explain the three distributions of transition frequencies; hence, sequential dependencies existed in the behavior sequences. Therefore, part of the information contributed by a given component in a sequence was shared with the following component. Since this information applies in common to the specification of both the preceding and the following component, it has to be subtracted from the uncertainty per component in order to yield the net information added by each component in the sequence (or the minimal amount of information necessary to specify the next component).

This is achieved by estimating the uncertainty associated with the two-part system or the joint occurrence of a behavior x and a following behavior y. The joint uncertainty $H(x,y)$ is defined as:

$$H(x,y) = -\sum_{ij} p(i,j)\log_2 p(i,j),$$

where the $p(i,j)$ are the probabilities associated with the joint occurrence of ($x=i$ and $y=j$). The joint uncertainty is compared to the sum of the individual uncertainties $H(x)$ and $H(y)$. (In an analysis of a stochastic series of events $H(x)$ is, of course, equal to $H(y)$ except for end effects, since each behavior but the first and the last one is considered twice: as a following act y and a preceding act x.) The deficit of the joint uncertainty $H(x,y)$ as compared

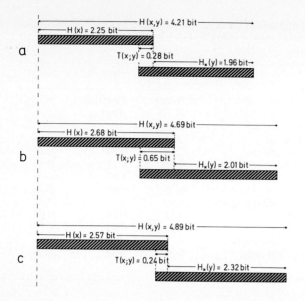

Fig. 4.21. Information theory analysis of the behavior displayed by an experimental *Gnathonemus petersii* ($n=7$) in the presence of *a* an intact conspecific, *b* an electrically "silent" conspecific, or *c* a *Brienomyrus niger* (Kramer 1976a)

with the sum of the individual uncertainties $H(x)$ and $H(y)$ is the amount of information $T(x;y)$ which is shared by two successive components:

$$T(x;y) = H(x) + H(y) - H(x,y).$$

The *T*-measure is a measure of the internal constraint within the system leading to an association between x and y. This is also given by the coefficient of constraint $D(x;y) = T(x;y)/H(x)$ (see Attneave 1974; and Table 4.1) which measures the average information which is in common with x and y, expressed as a portion of the entropy of x. The coefficient of constraint therefore varies between 0 (x and y totally unrelated) and 1 (total specification of y by x).

The strongest relatedness between a preceding and a following act was found in the experiment with the silent conspecific ($T(x;y) = 0.65$ bit; Table 4.1, Fig. 4.21) as compared with the experiments with the intact conspecific (0.28 bit) and *B. niger* (0.24 bit). These three *T*-values are all highly significant ($P < 0.001$), as shown by the χ^2-analysis mentioned above.

The average uncertainty $H_x(y)$ concerning the following event y, given that the preceding event x is known, can be calculated as the weighted average of the individual conditional uncertainties for each act x_i:

$$H_x(y) = -\sum_i p(i) \sum_j p_i(j) \log_2 p_i(j),$$

where the $p_i(j)$ are the conditional probabilities of $(y=j$ if $x=i)$, or $H_x(y) = H(x,y) = H(x)$, which is equivalent to $H_x(y) = H(y) - T(x;y)$.

This information measure gives the net amount of uncertainty per component which remains after having subtracted the gain of certainty about y from observing x (last equation).

Table 4.1 shows rather similar values for the average conditional uncertainties $H_x(y)$ for the experiments with the silent and the intact conspecifics

Table 4.2. Information theory analysis of the behavior displayed by *Gnathonemus petersii* in the presence of one of three types of stimulus fish (*cs* conspecific, or *Brienomyrus niger*) (Kramer 1976a)

Preceding behavior	Conditional uncertainty $H_i(y)$ in bit of the next behavior given that the preceding behavior is known Experiment with		
	Intact cs	Electrically silent cs	B.niger
Butt	2.04	2.07	2.44
Approach	1.94	2.00	2.32
Stop	2.05	1.96	2.06
Swimming	2.22	2.55	2.46
Swimming backwards	2.01	2.10	2.57
Probing	1.97	1.67	2.20
Hiding	1.77	1.45	1.99
Other acts	2.24	1.86	2.21
Average conditional uncertainty $H_x(y)$ in bit (ambiguity)	1.96	2.01	2.32

(2.01 and 1.96 bit, respectively); in the *B. niger* experiment a considerably higher value (2.32 bit) was found, which means that the average uncertainty about the following event, given that the preceding event was known, was greater.

Table 4.2 also gives the individual conditional uncertainties for each act. In each set, there was one which did not (or did not greatly) reduce $H_x(y)$ as compared with $H(x)$. For example, from observing swimming backwards in the experiment with *B. niger*, no certainty was gained about the following act, since the uncertainty derived from the frequencies of the components alone was no greater [$H_{SwBw}(y) = 2.57$ bit, $H(x) = 2.57$ bit]. All the other conditional entropies in each set were smaller than $H(x)$, however, and contributed to the reduction of $H_x(y)$ as compared with $H(x)$, according to their size and weight.

The redundancies of the behavioral sequences $(C_x(y) = 1 - H_x(y)/\log_2 n)$ found in the experiments with the intact conspecific, the silent conspecific, and *B. niger* thus rose from 25%, 11%, and 14% to 35%, 33%, and 23%, respectively, by considering first-order transitions. Thus, the greatest reduction of uncertainty by observing preceding acts was found in the experiment with the intact conspecific, because the *T*-value was highest.

A silent conspecific was less effective in eliciting a change from socially irrelevant (swimming, stop, etc.) to socially significant (agonistic) behavior in *G. petersii* than the two electrically intact stimulus fish. *B. niger* was as effective (or even more effective) in evoking attacks by the resident *G. petersii* as a conspecific, although the inter-EOD interval patterns of both species are quite distinct (see Sects. 3.2.1.1, 4.1.1.1, and 4.2.1.2). Therefore, the hypothesis could be advanced that all that matters for a resident *G. petersii* is whether the stimulus fish discharges or not, irrespective of the finer detail of the pulse patterns displayed. The two species' EODs are of similar waveform and frequency content; see Section 3.1.1.1.

This notion cannot, however, be maintained, because for every preceding behavior displayed by the experimental subjects, the distribution of following acts was significantly different for the three types of stimulus fish used.

The certainty $T(x; y)$ about the following act, gained from observing an experimental subject's preceding act, was only slightly different in the experiments with the intact conspecific and *B. niger* as stimulus fish (Table 4.1; Fig. 4.21). However, the remaining conditional uncertainty $H_x(y)$ about the following behavior was substantially greater [the redundancy $Cx(y)$ considerably lower] in the heterospecific experiment than in the two other series of experiments. Thus it was *B. niger* which elicited the most unpredictable behavior in the experimental animals. In the presence of a silent or an intact conspecific, the redundancies of the behavior sequences exhibited by the experimental animals were rather similar (33% and 35%, respectively), although there are important differences in the structure of the underlying information system. With an intact conspecific as a stimulus fish, the experimental subjects displayed a rather stereotyped behavior, resulting in a low $H(x)$. The silent conspecific elicited a much more diversified behavior, resulting in a high $H(x)$. The extent to which the following behavior y was specified by the nature of the preceding behavior x was small in the intact conspecific experiment; this extent was considerably greater in the silent conspecific experiment. Therefore, the conditional uncertainties $H_x(y)$ about the following behavior are similar.

In conclusion, the electric signals of both the conspecific and the heterospecific stimulus fish reduced the amount of sequential constraint in the overt behavior displayed by the experimental subjects by more than one-half, as compared with the silent conspecific experiment. A low degree of internal constraint in the behavior-producing system of both partners is a prerequisite for successful communication, since sufficient freedom of choice is necessary in order to respond to one's partner's unpredictable behavior. Another prerequisite is a high degree of redundancy in the behavior sequence, to allow time for the synchronization of partners in initially widely different behavioral and perhaps physiological states, and for the establishment of a hierarchy.

In the presence of an intact conspecific, the experimental fish displayed a highly redundant (in fact, the most redundant) behavioral sequence *in spite of* a very low amount of internal constraint. In the two other experiments, either redundancy was low (as in the *B. niger* experiment), or internal constraint high (silent conspecific experiment). Therefore, it is concluded that conspecific electrical signals transmit specific information to a receiver fish, which translate into overt behavior. The role of these signals may be especially important during reproduction, when it is crucial that the behavior of mates be synchronized in these nocturnal animals (see Sect. 4.1.1.1).

4.2.1.2 Playback of EOD Interval Patterns

Any investigation of electrocommunication in mormyrids would be incomplete without studying the effect of EOD patterns, which vary so greatly

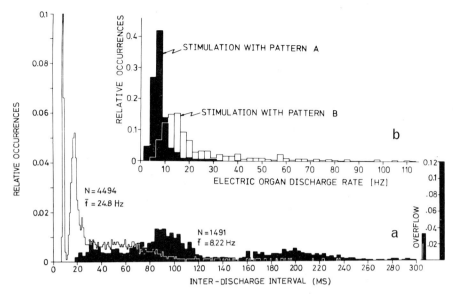

Fig. 4.22. EOD patterns used for 3-min stimulation tests of the experimental *Gnathonemus petersii* via a dipole model. *a* EOD interval histograms. Stimulus pattern A (*black histogram*, 3-ms bins) is the discharge activity of an isolated and resting *G. petersii*, while stimulus pattern B (*white histogram*, 1-ms bins) shows that of an attacking and charging fish. *Black ordinate* is for black histogram. *N* Total number of intervals; *f* mean discharge rate. *b* Histogram of EOD rates, as determined by counting the number of EODs during successive 0.2-s periods (same EOD sequences as shown in a). Note that rest pattern A shows a wider range of EOD intervals, but narrower range of EOD rates as compared with attack pattern B. *Abscissa* EOD rate (2.5-Hz bins); *ordinate* relative occurrences of specific EOD rates (Kramer 1979)

depending on behavioral context, on electric and motor behavior by playback. Different interval patterns ("signals") could code for different messages, translating into a variety of motor behaviors of a receiver fish. This hypothesis was tested by selecting two distinct EOD patterns, that of rest (A) and that of attack (B). These patterns were played back to the experimental subjects, the motor behavior of which was filmed by a video system and the EODs recorded on magnetic tape.

General features of the discharge patterns chosen for playback were described earlier: for patterns displayed by resting *Gnathonemus petersii*, see Section 3.2.1.1 (Figs. 3.25–3.27), those of aggressive *G. petersii* attacking and chasing an intruder, see Sect. 4.1.1.1 (Figs. 4.2–4.4, 4.6). The two EOD samples selected for the playback experiments (Fig. 4.22) fit that description.

Rest pattern A is a broad distribution of EOD intervals with three weak modes. Attack pattern B shows a bimodal EOD interval distribution with a very sharp gap separating the two modes, which is characteristic of an aggressive *G. petersii*. The mean discharge rates were 8 Hz for pattern A and 25 Hz for pattern B. Pattern A is confined to a narrow spectrum of EOD rates between 2.5 and 40 Hz, with most occurrences at 7.5 Hz (as determined by successive 0.2 s samples, during which EODs were counted). The range of pulse rates seen in pattern B is considerably larger, 5–112.5 Hz. The discharge rate measured most often was 15 Hz (Fig. 4.22b).

There were also differences in the sequence of pulse rates. In rest pattern A, pulse rate remained, on average, above or below the mean for ten successive observations (that is, 2.0 s) since nine autocorrelation coefficients were significantly positive; after that time no correlation was found (coefficients were not significantly different from zero). (For a short introduction into autocorrelation, see Sect. 3.2.1.1.) Superimposed on this simple kind of discharge rate regulation was an alternating discharge rate change between adjacent observation periods (that is, from one 0.2-s period to the next), since the second correlation coefficient was higher than the first one. The cause for this was an alternating EOD interval-length regulation between adjacent EOD in-

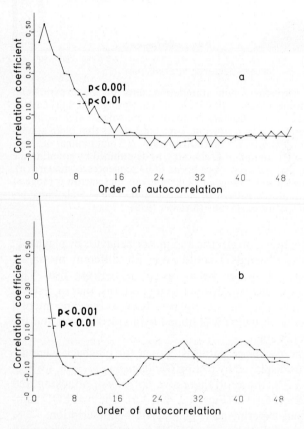

Fig. 4.23. Autocorrelation analysis of the pulse rates of rest and attack patterns A (*a*) and B (*b*) of a *Gnathonemus petersii* that were used for stimulation tests. Note that the autocorrelation function of attack pattern B starts with a high coefficient but declines more steeply than does that for rest pattern A. The positive pulse rate correlation of attack pattern B dies out within 0.8 s (3 significant coefficients, $P<0.001$), while pulse rates of rest pattern A are positively correlated for 2 s (9 significant coefficients). Superimposed on this is a short-term (0.2 s) pattern of alternating pulse rate in *a*, since the second coefficient is higher than the first (Kramer 1979)

tervals (see Fig. 3.26). So the discharge rate of the next observation but one was more similar to the first observation than the one between these two observations (Fig. 4.23).

In attack pattern B, pulse rates remained above or below the mean for a considerably shorter time only. After four observation periods (that is, 0.8 s), pulse rate was no longer correlated with respect to the first observation, since only three coefficients were significantly different from zero. Contrary to pattern A, pulse rates of adjacent observation periods were very strongly correlated in pattern B, since the autocorrelation function started at a very high coefficient of 0.76 (Fig. 4.23).

Attack pattern B fluctuated through a spectrum of discharge rates much more rapidly than did rest pattern A, as shown by Fourier analysis of the sequence of discharge rates. In attack pattern B the frequency of peak amplitude of these fluctuations was 0.23 Hz, and only 0.041 Hz in rest pattern A.

The most frequent reponse the experimental fish showed to both patterns was a bodily 'startle response' (or, maybe, movement of intention to leave the shelter), accompanied by a transient discharge rate increase, and followed by a retreat back into the shelter (Fig. 4.24a). On performing several of these responses, fish would sometimes leave their hiding places and attack the dipole model (Fig. 4.24b), usually followed by a lateral display (Fig. 4.24c) with high discharge rate (such as shown in Fig. 4.6). Other behaviors indicating high excitement were also observed during stimulation, such as swimming restlessly about the aquarium, probing the dipole model intensely with the chin appendage for more or less sustained periods of time, and swimming in rapid bouts near the bottom of the aquarium. All of these behaviors stopped when the 3-min stimulation period was over.

In every paired observation (pattern A vs B), the rate of startle reponses per second, displayed by a hiding fish, was higher during stimulation with attack pattern B compared with rest pattern A. The finding was significant individually for five out of seven fish ($P<0.05$, two-tailed randomization test for matched pairs); in two fish with similar results the data were insufficient for statistical analysis (because these fish tended to attack the dipole model immediately after stimulation onset). Pooling of the data of all fish yielded a $P \ll 0.001$ (Fig. 4.25).

Except for two fish which never left their shelters and therefore did not attack the dipole model, all other animals (that is, five) did so either more often or exclusively, during stimulation with attack pattern B as compared with rest pattern A ($P<0.05$; two-tailed randomization test for matched pairs).

Associated with attacks on the dipole model, high discharge rates were displayed which did not differ from those occurring in agonistic encounters between two fish. Likewise, the fish displayed very strong Preferred Latency Responses (PLRs; of up to 36% of all EODs within the preferred latency range, 9–13 ms), with both stimulus patterns. (The PLR had been observed to occur in agonistic context in *G. petersii*, see Sect. 4.1.1.1, and Fig. 4.9).

Very similar EOD rate increases (to about 40 Hz, starting from a low rest value <10 Hz) were observed at the onset of both stimulus patterns (Fig. 4.26, rest; and Fig. 4.27, attack).

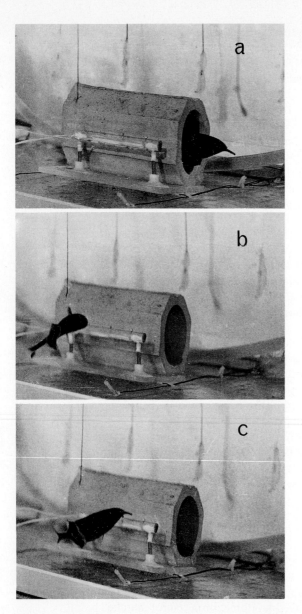

Fig. 4.24. Motor responses of *Gnathonemus petersii* on playbacks of social signals via a dipole. *a* Startle responses (or movements of intention to leave the shelter) were observed significantly more often during stimulation with attack pattern B compared with rest pattern A. A startle reponse was a rapid forward movement followed by retreat during which the fish never totally left its porous-pot hiding place. Startle responses were always accompanied by a transient EOD rate increase. *b* Attack responses on the dipole model were observed significantly more often during stimulation with attack pattern B. *c* Subsequent Lateral Displays were accompanied by high discharge rates, as was also observed in social agonistic encounters (Kramer 1979)

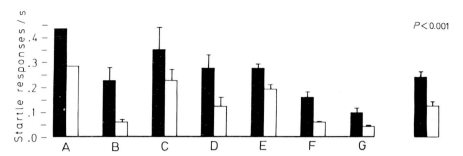

Fig. 4.25. Startle response rates displayed by seven *Gnathonemus petersii* (*A–G*) in their daytime shelters during stimulation with rest pattern A *(white columns)* and attack pattern B *(black)*, via an electric dipole model. Means +1 SE, separately for each fish (no SE for fish A because of lack of data), and pooled *(right pair of columns, $P \ll 0.001$)*. When stimulated with attack pattern B, each fish displayed a higher rate of startle responses as compared with rest pattern A (After Kramer 1979)

After about 10 s, the responses to both stimulus patterns began to differ. Although in both instances EOD rates remained high compared to the time before stimulation onset, modulation of the fishes' EOD rate sequences was of stronger amplitude and higher frequency during stimulation with attack pattern B than with rest pattern A. Maximum EOD rates displayed by the experimental fish were much higher during stimulation with the B pattern compared with the A pattern. Specifically interesting is the observation that fish "imitated" (or responded in a similar fashion to) pulse rate increases of the dipole, especially when playing back pattern B which contained many high discharge rates (that are associated with attack and lateral display during agonistic encounters). In the example of Fig. 4.27, 15 such EOD rate increases (>60 Hz) occurred during stimulation with the B pattern (180 s), in response to an increase of the dipole pulse rate. There were only two such responses, which were also weaker (>40 Hz), to the A pattern (Fig. 4.26).

Auto- and cross-correlation methods are useful to characterize these differences between the responses to the A and B patterns. The analysis of the EOD rates of stimulated fish invariably yielded positive autocorrelation coefficients for adjacent data points, with a decline to zero as more data points separated the two points considered (Fig. 4.28).

However, the EOD activities of the experimental subjects showed significant differences, depending on the type of stimulation pattern used. The correlation between adjacent data points of the EOD rate sequence was stronger when the fish were stimulated with rest pattern A (average correlation of 0.67) compared with attack pattern B (0.61, difference significant with $P<0.02$; Figs. 4.28 and 4.29). Also, the number of correlation coefficients significantly different from zero ($P<0.001$) was greater during rest pattern stimulation (on the average 9.7) than during attack pattern stimulation (on the average 6.2, difference significant, $P<0.01$; Fig. 4.29).

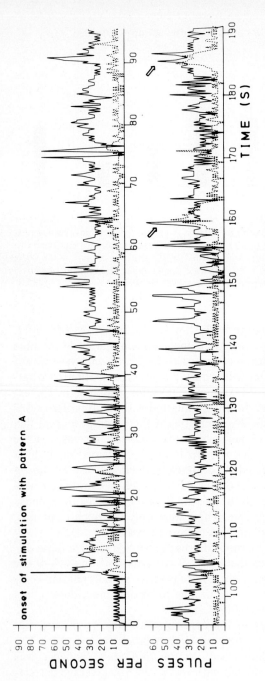

Fig. 4.26. Concurrent rates of the EODs of a *Gnathonemus petersii* (*line*), and of the stimulus pulses of a dipole model (*broken line*) playing back rest pattern A. *Abscissa* time (s); *ordinate* pulse rates measured during successive 0.2-s time intervals (expressed in Hz). The lower part is the continuation of the upper. *White arrows* indicate two instances where an increase of the stimulus pulse rate coincided with an EOD rate increase. Note that these instances were rare compared with Fig. 4.27 (where the attack pattern B was used for stimulation). However, even with the present rest pattern stimulation, *G. petersii*'s low resting EOD rate increased markedly on stimulation onset (Kramer 1979)

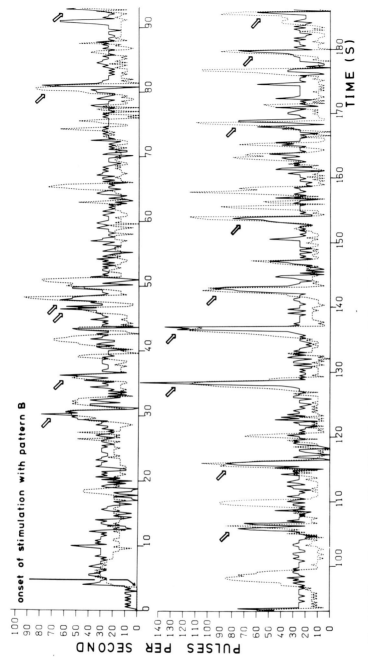

Fig. 4.27. Shows the same as Fig. 4.26 but for stimulus pattern B (attack). Note that *Gnathonemus petersii* responded to an increase of stimulus pulse rate by an EOD rate increase (*white arrows*) much more often in the present experiment, compared to that using rest pattern A (see Fig. 4.26; Kramer 1979)

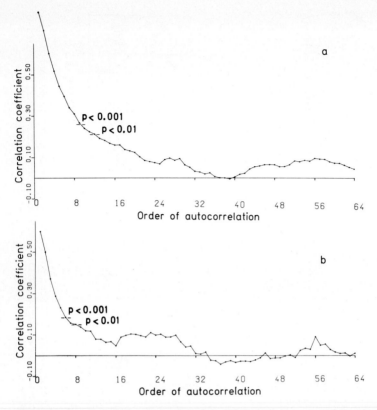

Fig. 4.28. Autocorrelation examples of the EOD rate responses of *Gnathonemus petersii* to *a* stimulation with rest pattern A and *b* stimulation with attack pattern B. Note that the autocorrelation function of the EOD responses start with a higher coefficient in *a* than in *b*. Discharge rate correlation died out at a greater lag or order of autocorrelation in *a* (2.0 s) than in *b* (1.4 s). Data from same fish and day (Kramer 1979)

Other differences were also found. The EOD rates measured most often (as determined from pulse rate histograms such as shown in Fig. 4.22b) were significantly lower during rest pattern stimulation (average 12.3 Hz) compared with attack pattern stimulation (average 16.2 Hz; $P<0.001$, Fig. 4.29). Similar differences (18.2 Hz and 21.3 Hz, respectively) were obtained for the mean discharge rates ($P<0.001$). The span of pulse rate histograms (from lowest to highest rates) comprised an average 47 Hz during rest pattern stimulation, and an average of 56 Hz during attack pattern stimulation ($P<0.05$, Fig. 4.29).

The periods of time for which similar discharge rates were maintained were longer during rest than during attack pattern stimulation, as shown by a greater number of positive correlation coefficients (significantly different from zero) during rest compared with attack pattern stimulation. Spectral analysis of the time sequences of EOD rates confirmed that the fishes' discharge rates fluctuated at different frequencies, depending on the stimulus pattern. The

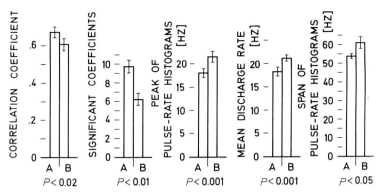

Fig. 4.29. Statistical comparison of autocorrelation and pulse rate parameters of the EOD activities displayed by seven *Gnathonemus petersii*, during stimulation with rest pattern A (*left columns*) and attack pattern B (*right columns*). The means ($n=23$) are shown with their ± 1 standard error ranges. Note that the correlation between adjacent data points was stronger, and the number of coefficients significantly different from zero ($P<0.001$) greater, during rest pattern stimulation compared with attack pattern stimulation. The modes of the pulse rate histograms were significantly different, and also the mean EOD rates differed significantly between the two experiments. The widths of the pulse rate histograms (that is, the span from the lowest to the highest rates) were greater during attack pattern stimulation than during rest pattern stimulation. Two-tailed *P*-values were calculated by the paired *t*-test (except the second pair of columns from the left: Wilcoxon matched-pairs signed-ranks test; Kramer 1979)

means of the fluctuation frequencies of peak amplitude were near 0.02 Hz during rest pattern stimulation, and approximately 0.07 Hz during attack pattern stimulation ($P<0.01$). The highest frequencies of one-third peak amplitude were on the average 0.16 Hz during rest, and 0.48 Hz during attack pattern stimulation.

Several significant differences between the responses of the fish to the two stimulation patterns were also detected by cross-correlation of the EOD activities to the stimulus patterns. EOD responses of the fish to the stimulus patterns were weaker during rest pattern stimulation (average maximal cross-correlation of 0.2) than during attack pattern stimulation (average maximal cross-correlation of 0.33; $P<0.001$; Figs. 4.30, 4.31).

However, the number of significant ($P<0.001$) cross-correlation coefficients was greater during rest pattern stimulation than during attack pattern stimulation ($P<0.01$), irrespective of whether positive coefficients only, or coefficients of either sign were considered.

Cross-correlation was not only weaker during rest pattern stimulation compared with attack pattern stimulation, but the lag of maximal cross-correlation was also greater (that is, the maximal effect of the stimulus pattern on the EOD sequence was delayed relative to the results with the attack pattern). During rest pattern stimulation an average lag near 2.6 s was found; during attack pattern stimulation the average lag was only 0.8 s ($P<0.01$, Fig. 4.31).

The results clearly show that both the motor as well as the electric responses of the receiver fish were significantly different for both stimulation

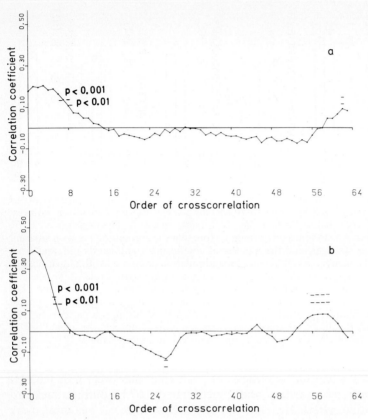

Fig. 4.30. Cross-correlogram examples of the EOD rate responses of a *Gnathonemus petersii* to stimulation with *a* rest pattern A and *b* attack pattern B. Note that the strongest cross-correlation in *a* was weaker and occurred at a greater lag (0.6 s), or order of cross-correlation, than in *b* (0.2 s). However, significant cross-correlation ($P < 0.001$) was maintained for a longer period of time in *a* (1.4 s) than in *b* (0.8 s). Data from same fish and day (Kramer 1979)

patterns in many ways. It can be concluded that at least two specific time patterns of pulses encode different messages in the intraspecific communication system of *G. petersii*.

The relationships between stimulus and response patterns are, however, far from simple. While the pulse rate histogram data might partly be explained by a very simple "imitation model" (a stimulus pattern with a broad distribution of pulse rates, shown in Fig. 4.22, elicited a similarly broad response distribution of pulse rates), no such hypothesis can be maintained for part of the autocorrelation results of the same data. Autocorrelation between adjacent EOD rate measurements was stronger during rest pattern stimulation compared with attack pattern stimulation (Figs. 4.28, 4.29), although for the two stimulation patterns the reverse results were found (Fig. 4.23). Despite this, the periods of time of positive correlations of the EOD response patterns were longer during rest pattern stimulation (2 s) than during attack

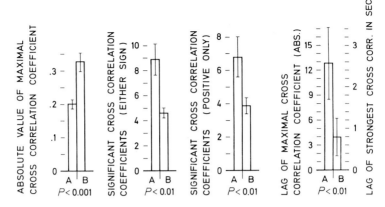

Fig. 4.31. Statistical analysis of the cross-correlations from stimulus (rest pattern A, *left columns*, and attack pattern B, *right columns*) to the EOD response patterns of seven *Gnathonemus petersii*. Means ($n=23$) are shown with their ± 1 SE. Note that the maximal cross-correlation coefficient is lower, but the number of coefficients significantly different from zero ($P<0.001$) greater, during rest pattern stimulation compared with attack pattern stimulation. The lag of strongest cross-correlation was approximately 2.6 s during rest pattern stimulation, and 0.8 s during attack pattern stimulation. Two-tailed *P*-values were calculated according to the Wilcoxon matched-pairs signed-ranks test (except the left pair of columns: paired *t*-test; Kramer 1979)

pattern stimulation (1.2 s, Fig. 4.29), and thus approximately matched properties of the stimulus patterns (rest pattern: 2 s; attack pattern: 0.8 s, Fig. 4.23). In a similar fashion, high fluctuation frequencies of discharge rate changes were elicited by the attack pattern with similar property. Surprisingly, the attack pattern had a much stronger influence on the reponse pattern than did the rest pattern, as measured by cross-correlation (Figs. 4.30 and 4.31). In addition, this stronger correlation occurred at a shorter lag. This shows that rapid discharge rate changes may be among the important features that are detected by the fish and that influence arousal and motivational state.

Much more work remains to be done in order to understand electrical communication in *G. petersii*; for example, the signal value of other EOD displays should also be determined, such as those occurring during "swimming", "fleeing" (Sects. 3.2.1.1. and 4.1.1.1), and reproduction (yet unknown in *G. petersii*; but see Sect. 4.1.1.1 for *Pollimyrus isidori*). The function of certain EOD displays could also depend on context, as seen in *P. isidori*. The characteristic features of displays could be determined experimentally by artificially modifying natural EOD patterns.

4.2.1.3 Species Recognition by EOD Interval Pattern
(B. Kramer and H. Lücker)

Moller and Serrier (1986) showed that stimulus fish of four species, confined in a sector at the periphery of a large tank, strongly attracted freely moving

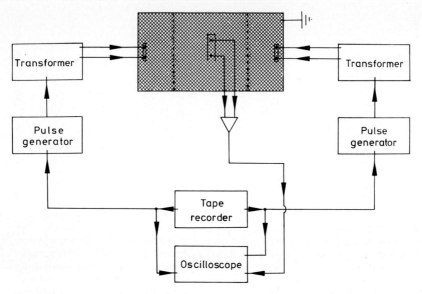

Fig. 4.32. Apparatus for preference tests using simultaneously two inter-EOD interval patterns. An experimental fish (*Petrocephalus bovei*) was hiding in a porous pot (length, 17 cm; inner diameter, 4 cm) in the center of the tank (conductivity, 350 µS/cm; 27 ± 1°C). Fish were simultaneously stimulated via two dipoles (3 mV/cm measured inside the porous pot) with two different EOD sequences stored on the two tracks of a magnetic tape (Revox A77 tape-recorder). Each EOD stored on the tape triggered a single 3-kHz sine-wave pulse from a function-generator (one for each EOD sequence and dipole); the dipoles were isolated from ground by symmetrical transformers. With neither reward nor punishment, preference for one of the two pulse patterns was measured as the time an experimental fish spent near the dipoles (within an area marked by *dashed lines*, outside its central "home range"). Aquarium length, 75 cm; width, 40 cm. An experimental design of randomized blocks was used to compensate for sidedness

conspecifics (*Brienomyrus niger, Gnathonemus petersii, Marcusenius cyprinoides, Pollimyrus isidori*). Confined heterospecific stimulus fish proved less attractive than conspecific ones in most cases, or not attractive at all (depending on the species pairing; see also Serrier and Moller 1981). In these observations during daylight hours the sensory modality involved was not controlled (the electrical, the visual, the mechanical, the auditory, and perhaps chemical, could all be involved). However, the results of denervation experiments of the electric organ (Moller 1976; Kramer 1976a) suggested the great importance of electrical signals also here.

Do mormyrids recognize conspecifics from their EOD interval time patterns when all other cues are excluded? Playback experiments were designed to find an answer to this question (preliminary report, Lücker 1983).

In a randomized choice experiment with neither reward nor punishment, isolated resident *Petrocephalus bovei* were simultaneously stimulated with two discharge patterns stored on the parallel tracks of a magnetic tape (experimental setup, see Fig. 4.32).

In most experiments, one of the two patterns used for play-back was a conspecific pattern, the other pattern being either a similar one recorded from another individual, or one recorded from a different species (*P. isidori* or *B. niger*). Two types of patterns from each species were used: resting patterns (previously recorded during the diurnal inactive period, and only when isolated fish were completely immobile), and swimming patterns (previously recorded during the nocturnal period of activity, and only when fish were spontaneously swimming about, with no prior disturbance). Short sections are shown in Figs. 4.33 and 4.34.

Prior to each experiment two patterns were chosen according to a randomized blocks design (Cochran and Cox 1957), in order to compensate for possible sidedness and fluctuations of internal state (motivational changes). Restricted random permutation of pattern pairings and assignment of sides was achieved as follows (separately for resting and swimming pat-

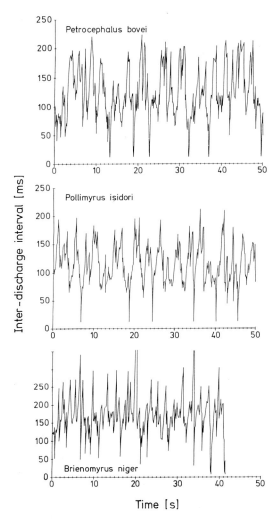

Fig. 4.33. Resting discharge activity of isolated mormyrids of three species, recorded during day-light hours when fish were completely immobile. Sequential representations of discharge activities. *Ordinates* duration of individual inter-discharge intervals [ms]; *abscissae* time of occurrence of each discharge [s]. *Top* Petrocephalus bovei; middle Pollimyrus isidori; bottom Brienomyrus niger

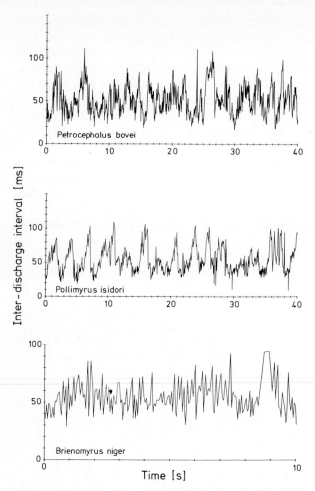

Fig. 4.34. Discharge activity of isolated mormyrids of three species, recorded during the nocturnal period of spontaneous swimming (no rest periods included). As Fig. 4.33. *Top Petrocephalus bovei; middle Pollimyrus isidori; bottom Brienomyrus niger*

terns): First, one of four pairs of discharge patterns was randomly selected (*P. bovei* No. 1 – *P. bovei* No. 2; *P. bovei* – *P. isidori*; *P. bovei* – *B. niger*; *P. isidori* – *B. niger*); second, the selected pair of patterns was randomly assigned to the right and left dipoles and the experiment carried out. Step 3 was the repetition of that experiment with sides reversed. Step 4 was identical with step 1, with the difference that the pair(s) of patterns already used was (were) no longer considered until every pattern pairing had been used once.

Spontaneous preference was measured as the time the experimental *P. bovei* ($n=9$) spent near each dipole. The duration of each experiment ($n=780$) was 10 min, the inter-trial interval was at least 15 min. Experiments were performed during daylight hours (L:D, 12:12); stimulation was turned on only when fish were hiding.

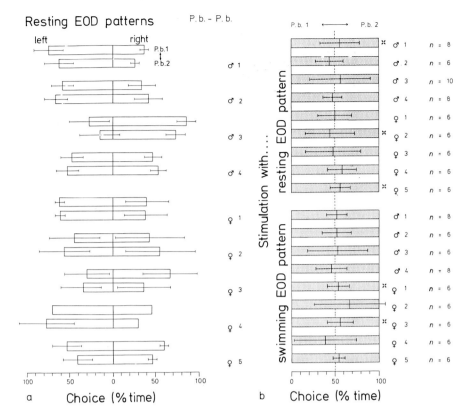

Fig. 4.35. Preference test for two conspecific EOD interval patterns, by simultaneous playback via two dipoles, one on each side of an isolated *Petrocephalus bovei* ($n=9$) hiding in a centered porous pot (see Fig. 4.32), shown individually for all experimental subjects (4 males and 5 females). *a* Test for sidedness using two similar, but not identical, resting patterns previously recorded from two *P. bovei* (P.b.1, *upper bars*; P.b.2, *lower bars*). The *abscissa* shows the average time (± 1 SD) an experimental fish stayed close to the *right* or the *left* dipole as a percentage of the total time it spent outside its central "home range", in a specified area near a dipole. Note marked individual side preferences but small differences between patterns. *b Top half* shows the same as *a* but data normalized for showing differences between patterns rather than sides (data pooled for both sides). Note that only three fish showed a statistically significant preference for one of the two *P. bovei* resting patterns (* $P<0.05$; without "agreement" on one pattern). *Bottom half* shows the same as top half except that the two stimulus patterns were similar, but not identical, swimming patterns previously recorded from two *P. bovei*. Only two fish of nine showed a statistically significant preference for one pattern (as indicated)

Most fish showed a preference for the right or the left dipole even when two similar conspecific patterns were played back (Fig. 4.35 A).

Female No. 1, for example, tended to move to and stay near the left dipole indiscriminate of the stimulus pattern. It therefore received a "pattern preference score" of close to (and not significantly different from) 50% (Fig. 4.35 B); that is, it did not show a preference for either stimulus pattern. The

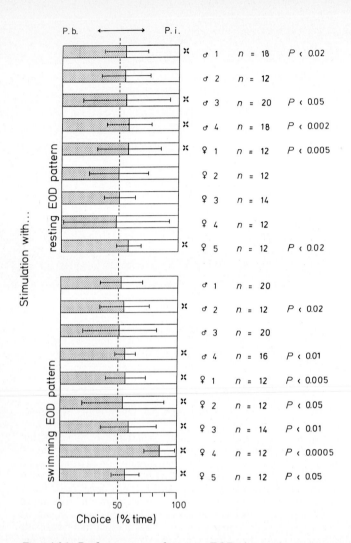

Fig. 4.36. Preference test for two EOD interval patterns (one conspecific), by simultaneous play-back via two dipoles, one on each side of an isolated *Petrocephalus bovei* (*n*=9) hiding in a centered porous pot (see Fig. 4.32). The *abscissa* shows the average time (± 1 SD) an experimental fish spent near a dipole playing back *P. bovei* (*stippled*) or *Pollimyrus isidori* patterns (*open bars*), as a percentage of the total time a fish was near a dipole (within a specified area, see Fig. 4.32). Randomized blocks design of experiments without reward or punishment. *Top half* stimulus patterns were resting patterns. *Lower half* stimulus patterns were swimming patterns. Note that five *P. bovei* significantly preferred their conspecific resting pattern to that of *P. isidori* (*; significance level as indicated), while no fish preferred *P. isidori* resting or swimming patterns to those of *P. bovei*. Seven *P. bovei* significantly preferred a dipole playing back a conspecific swimming pattern to one playing back a *P. isidori* swimming pattern

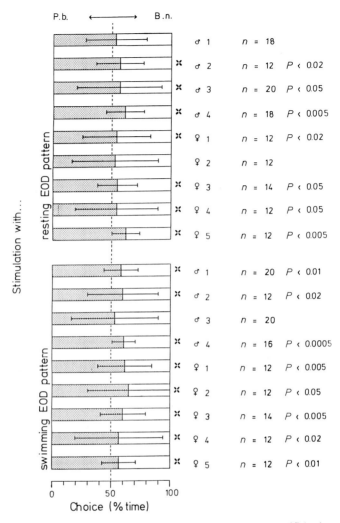

Fig. 4.37. Preference test for two EOD interval patterns (one conspecific), by simultaneous playback via two dipoles (as in Fig. 4.36). The experimental *Petrocephalus bovei* (n=9) could choose between conspecific EOD patterns (*stippled*) and those from *Brienomyrus niger* (*open bars*). Top half resting patterns; *lower half* swimming patterns. Note that seven *P. bovei* showed a significant preference for their conspecific resting pattern (*, *P* level as indicated), while no fish preferred *B. niger* resting or swimming patterns. Eight *P. bovei* preferred a dipole playing back a conspecific swimming pattern to one playing back a *B. niger* swimming pattern

pattern preference score was calculated by pooling the times spent near each dipole (separately for each pattern) and finding the percentage of the total time it had spent near a dipole. Three out of nine fish did, however, show a preference for one of the two conspecific resting patterns ($P<0.05$; Wilcoxon matched-pairs signed-ranks test), although there was no agreement among these fish about which pattern was the more attractive. Only two of nine fish had a preference for one of the two conspecific swimming patterns ($P<0.05$). It is unclear on the basis of which cues these fish preferred one pattern to the other.

Almost all *P. bovei* clearly preferred a conspecific swimming-pattern to one recorded from a *P. isidori* ($P<0.05$ in seven of nine fish) or a *B. niger* ($P<0.05$ in eight of nine fish) (Figs. 4.36 and 4.37). None of the experimental fish found one of the two heterospecific swimming patterns more attractive than the conspecific one.

This was also true for resting patterns: only conspecific patterns attracted *P. bovei* for significantly longer times than would be expected from chance ($P<0.05$ in five of nine fish in the *P. isidori* – *P. bovei* pattern pairing; $P<0.05$ in seven of nine fish in the *B. niger* – *P. bovei* pattern pairing) (Figs. 4.36 and 4.37).

Does *P. bovei* find *B. niger* patterns more (or less) attractive than *P. isidori* patterns? Both species' patterns seem to be equally attractive for most *P. bovei*, as only one (of nine) showed a significant ($P<0.05$) preference for the *B. niger* resting pattern compared with that of *P. isidori*; three *P. bovei* (of nine) showed a preference for one of the two heterospecific swimming patterns (without agreement as to which species' swimming pattern was the more attractive; Fig. 4.38).

Our *P. bovei* also markedly preferred conspecific to heterospecific pulse patterns in terms of their first choices when leaving their hiding tubes on stimulus onset. Both male and female *P. bovei* clearly preferred their conspecific resting and swimming patterns to the corresponding patterns of *B. niger* ($P<0.05$, sign test; Fig. 4.39).

The pairing of *P. bovei* patterns with those of *P. isidori* yielded similar results (except *P. bovei* males whose preference for their conspecific resting-pattern to that of *P. isidori* was not significant in terms of "first choices", although three of four males did show such a preference in terms of "time spent near a dipole"; Fig. 4.36).

The observation that there is no "first choice preference" of male or female *P. bovei* for one of two similar, but not identical, conspecific swimming patterns underlines the validity of these results (Fig. 4.39). Although males did show such a preference for one of two conspecific resting patterns ($P<0.05$), only one (of four) males showed a similar preference in terms of time spent near a dipole. Also, in terms of first choices, neither set of heterospecific pulse patterns was clearly more attractive than the other: male *P. bovei* were indiscriminate of the two heterospecific resting patterns, female *P. bovei* of the two heterospecific swimming patterns. However, male *P. bovei* preferred the *P. isidori* resting pattern to that of *B. niger* ($P<0.01$; a preference not present in terms of "time spent near a dipole") while female *P. bovei* preferred the swimming pattern of *B. niger* to that of *P. isidori* ($P<0.02$; a preference equally not observed in terms of "time spent near a dipole").

These experiments show a clear preference for conspecific resting and especially swimming patterns to the corresponding patterns from two other species in *P. bovei*. The only cues available to the fish were patterns of inter-pulse intervals; information about the species-characteristic EOD waveform was not

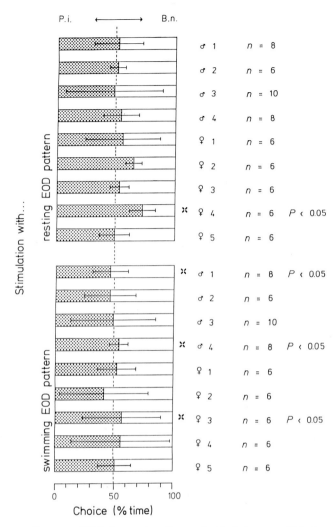

Fig. 4.38. Preference test for two EOD interval patterns (both heterospecific), by simultaneous playback via two dipoles (as in Fig. 4.36). The experimental *Petrocephalus bovei* (*n* = 9) could choose between two resting (*top half*) and two swimming patterns (*bottom half*), both from other species. *Stippled* The stimulus pattern was a *Pollimyrus isidori* pattern; *open bars* the stimulus pattern was a *Brienomyrus niger* pattern. Note that only one *P. bovei* showed a differential preference for one of the two resting patterns (that of *P. isidori*; *, *P* level as indicated), all other fish (8) were indiscriminate. In the swimming pattern experiment six fish did not show a significant preference for either of the two patterns, but three did (without "agreement")

present and hence is unnecessary for a correct choice (however, it may perhaps increase the likelihood of a correct choice). This discrimination was already seen in the fishes' first choices. We conclude (1) that the fish can recognize conspecific EOD interval patterns from those of other species, even when rather similar; (2) that it can do so even in the presence of the "noise" of the

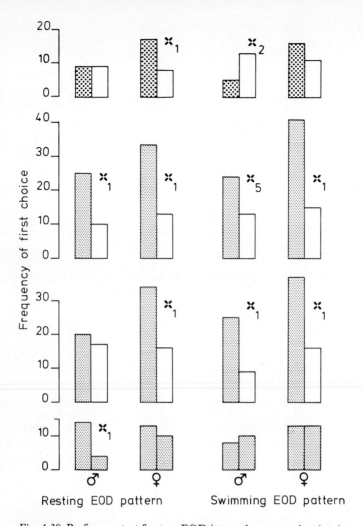

Fig. 4.39. Preference test for two EOD interval patterns, by simultaneous playback via two dipoles (as in Fig. 4.36). The *ordinates* show the fishes' (*Petrocephalus bovei*) frequency of first choices following stimulus onset, separately for males ($n=4$) and females ($n=5$; as indicated at the *bottom*). *Left half of figure* stimulus patterns were *resting* EOD patterns; *right half of figure* stimulus patterns were *swimming* EOD patterns (see *bottom*). *Fine stippling* stimulus patterns were conspecific *P. bovei* patterns. *Bottom row* simultaneous stimulation with two *P. bovei* resting patterns, and two *P. bovei* swimming patterns; similar, but not identical, patterns previously recorded from different individuals. Note that the experimental *P. bovei* did not prefer one conspecific resting or swimming pattern to the other (except in one comparison marked by an *: male *P. bovei* stimulated with two conspecific resting patterns; sign test). *Second row from bottom*, *P. bovei* vs *Pollimyrus isidori* patterns. Note that the fish preferred the conspecific patterns to those of *P. isidori* (although not significantly in the case of males stimulated with resting patterns; no *). *Third row from bottom*, *P. bovei* patterns vs *Brienomyrus niger* patterns. *P. bovei* of both sexes were attracted significantly more often by conspecific patterns than by heterospecific ones. *Top row* presentation of *P. isidori* (*coarse stippling*) and *B. niger* patterns (*open columns*) to the experimental *P. bovei*. Note that *P. bovei* males did not show any significant preference for one of the

discharges of another fish with very similar mean EOD rate and discharge interval distribution (see below).

P. bovei clearly discriminates the stimulus pulses it receives according to the spatial positions of the senders, otherwise one pattern would blur the other, and the fish would not be able to find the dipole playing back conspecific patterns already in their first choices. That dipole was chosen significantly more often than predicted by chance, in spite of random permutation of the roles of the left and the right dipoles. The mode of location of a dipole source was investigated in *Brienomyrus brachyistius* (Schluger and Hopkins 1987; see Sect. 2.3.2).

P. bovei's releasing mechanism for recognizing conspecific pulse patterns must be extremely good: The inter-pulse interval histograms of the three species' swimming patterns (see Figs. 3.24 and 3.28) that were actually chosen for the playback experiments were not significantly different from each other ($P > 0.2$, Kolmogoroff-Smirnoff test) and all had a mean discharge rate of 16–17 Hz.

Therefore, the species-characteristic differences must reside in the sequence of inter-pulse intervals of different durations. A comparison of interval trend plots shows this is the case (Figs. 4.33 and 4.34), although it is often difficult to recognize patterns in mormyrid EOD activity, because of an apparent random component superimposed or interspersed. A major problem is also our ignorance regarding which features we should look for. However, the fish obviously deal considerably better with this situation than we humans do, for in the present experiments they recognized their own species' patterns in their first choices significantly more often than predicted by chance. It is necessary, then, to look for species-characteristic and possibly species-specific EOD patterns lasting not more than a few seconds, because of the fishes' quick decision. The formidable power of visual pattern recognition humans are endowed with should help us to find such patterns. Such hypothetical patterns will be called "micropatterns" in the following paragraphs.

B. niger's swimming pattern differs markedly from that of the two other species by its strong tendency of adjacent intervals to alternate in duration (Fig. 4.34). This tendency also shows up in the autocorrelogram, by the second coefficient being higher than the first (Fig. 4.40). This alternating of (relatively) short with longer EOD intervals was only observed in fish swimming spontaneously at night, and not during swimming evoked by disturbance during the light period (Kramer 1976a). The alternating character may well represent a micropattern as defined above. Superimposed on this EOD activity seems to be a low-frequency (period around 3 s) modulation (gradual waxing and waning) of interval length, adding "width" to the interval histogram. (For a Fourier analysis of the spectrum of EOD rate fluctuations, see Kramer 1979.)

---

two heterospecific resting patterns, nor females for one of the two swimming patterns, although females preferred the *P. isidori* resting pattern to that of *B. niger*, for unknown reasons. This is offset by the males preferring the *B. niger* swimming pattern to that of *P. isidori*

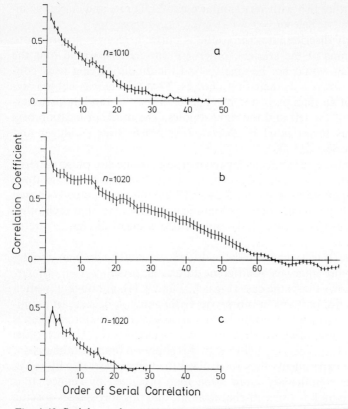

Fig. 4.40. Serial correlograms, or autocorrelation, of interpulse interval sequences of *a Petrocephalus bovei*; *b Pollimyrus isidori*; *c Brienomyrus niger*. Isolated fish during swimming, *a* and *c* during the night. Correlation coefficients that are statistically significantly different from zero, or no correlation are marked by *long vertical bar* ($P < 0.01$). Note that only *B. niger* shows a tendency of adjacent intervals to alternate in length, as shown by the second coefficient being higher than the first (*b*, Kramer 1978)

P. isidori's swimming pattern is considerably more complex (Fig. 4.34). A recurrent EOD pattern seems to consist of about 35–70 intervals and to last for about 2–4 s and will be called a micropattern. An arbitrary start for such a micropattern often seen is a long interval (about 100 ms) followed by a sharp discharge-rate increase in one or very few steps to an interval duration of about 50 ms. A series of intervals with a tendency of becoming still shorter (20–40 ms) follows before interval length increases again. The very high first correlation coefficient shows that neighboring intervals are, on average, very close in size, but there is a steady decline in similarity with increasing separation (Fig. 4.40).

P. bovei's swimming pattern shows fewer jumps from a low to a higher discharge rate in just one step compared with *P. isidori's* swimming pattern (Fig. 4.34). Otherwise the two species' patterns are rather similar, except for tendencies of intervals to change for some time, often being less clear in *P. bovei*. This is found when: (1) centered on the actual (moving) average there is a tendency of intervals to alternate; (2) the extreme intervals differ so little that the tendency to alternate masks any pattern.

The observation that of nine experimental *P. bovei*, eight discriminated the *B. niger* swimming pattern from their own species' pattern, compared with only seven of nine fish which discriminated the conspecific pattern from that of *P. isidori*, perhaps supports the notion that the differences between *B. niger's* and *P. bovei's* EOD patterns are greater than those between *P. isidori's* and *P. bovei's* patterns.

Micropatterns are not easily recognized from *B. niger's* resting-activity with its two modes and low mean discharge rate (4–8 Hz; for histogram, see Figs. 3.24 and 3.25), except occasional bursts of high discharge rate (usually up to about 12 intervals with the shortest interval around 14 ms) interrupting a lower discharge rate with intervals ranging from 120–250 ms (Fig. 4.33). Every few seconds there is either a single very long interval (greater than 250 ms), or a single relatively short interval (about 50 ms). These short intervals sometimes, but not always, terminate a trend, representing its minimum.

P. isidori's resting activity shows three modes, and the mean discharge-rate is relatively high (about 10 Hz; Figs. 3.24 and 3.25). Bursts interrupt a low discharge rate fairly regularly; they consist of about ten intervals as short as 11 ms (Fig. 4.33). These bursts thus resemble those displayed by *B. niger*. In contrast to *B. niger*, periods of low discharge rate exhibit micropatterns. These last for about 2–5 s and consist of 20–40 intervals. A group of medium-duration intervals tend to alternate with a group of long-duration intervals (on average, the second group is smaller). The transitions between both groups are brisk.

P. bovei's resting activity displays only two modes with a mean discharge rate of 8–9 Hz (Fig. 3.25). Bursts resemble those of *P. isidori* and *B. niger*, except for being shorter (4–6 intervals as short as 12 ms). Because of the similarity of all three species' burst patterns, it is highly unlikely that they convey species-specific information in the context of these experiments, except perhaps for their timing. A micropattern much more suitable for recognizing members of its own species in *P. bovei* would be a sequence of 25–35 intervals lasting for 2–4 s (Fig. 4.33). Starting from an interval of around 100 ms (representing the low discharge rate mode), two trends were seen. The first, leading to a burst by a continuous shortening of EOD intervals, is the one supposed to be of little use in species recognition (see above). The second, more interesting one, consists of a gradual shift to longer intervals, followed by a return to the original interval length. This represents a discharge-rate change from medium to low and back again, and was smoother than that seen in *P. isidori*, which tended to jump from medium to low discharge-rate levels. This difference might provide a species-specific cue in the *P. isidori* – *P. bovei* resting-pattern pairing.

Swimming patterns seem to be behaviorally more relevant than resting patterns that vanish with disturbance or movement of the fish. It is interesting to note that still more *P. bovei* (seven and eight of nine) recognized the conspecific swimming-pattern from the two heterospecific swimming patterns, compared with the results obtained with resting patterns (only five and seven of nine fish). This is remarkable in view of the high degree of similarity of all three species' swimming patterns in terms of inter-EOD interval histograms and mean discharge rates (differences not significant), which contrasts with the dissimilarity of the resting histograms (differences highly significant).

Several *P. bovei* kept together in a large tank tend to aggregate and school during the light period. Even individuals isolated for several months are active during light hours most of the time; they rarely remain immobile sufficiently long for obtaining at least a crude resting histogram (for which a rest period of about 3 min is required). In the Ivindo river (Belbenoit, pers. commun.) *P. bovei* have also been observed schooling during the day; isolated individuals were never found. Schooling fish are unlikely to discharge in resting-pattern fashion.

Although not tested directly, the above experiments (Figs. 4.36 and 4.37) point to the possibility that a conspecific swimming pattern is more attractive than a conspecific resting pattern in *P. bovei*, at least during the light period (and probably still more so at night, the natural period of activity). (In most pattern pairings, more fish were attracted by a dipole playing back swimming patterns compared with resting patterns, for a significantly longer time.) This presumed greater attractiveness of swimming patterns compared with resting patterns could, however, be reversed in males during the reproductive period. This is inferred from the reproductive behavior of *P. isidori* which has recently been elucidated (see Sect. 4.1.1.2). Male *P. isidori* construct nests and patrol their territories while females stay nearby, most of the time hiding completely immobile, and discharging in a resting pattern-like fashion (Bratton and Kramer 1989). It is during this time that resting patterns might serve a function in species and sex recognition (even individual recognition), and signal readiness to engage in courtship.

4.2.1.4 Discrimination of Inter-Pulse Intervals
(B. Kramer and U. Heinrich)

In the previous section of this chapter (4.2.1.3), *Petrocephalus bovei's* remarkable ability to recognize conspecific pulse patterns from those of two other mormyrid species was described. Paradoxically, discrimination was best when the stimulus pulse patterns presented were most similar (virtually identical) in terms of mean EOD rates and EOD interval histograms, as in the case of EOD patterns of swimming fish (Figs. 4.36 and 4.37). These swimming EOD patterns did, however, differ in typical interval sequences lasting a second or two, or micropatterns (Figs. 4.34, 4.40). Discrimination of interval patterns is only possible if fish can measure inter-pulse intervals precisely, a sensory capacity to date never investigated in weak electric fish.

The discrimination limen of *Pollimyrus isidori* for pulse intervals of different lengths was tested by using pulse trains of constant rate, a specific pulse rate being the rewarded stimulus S+, the S- being another which was associated with mild punishment. Fish were trained to discriminate between the two.

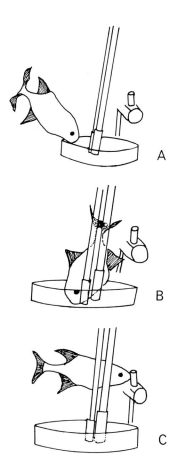

Fig. 4.41. A, B. Responses of a trained *Pollimyrus isidori* to the rewarded stimulus, S+, a pulse train of constant rate, presented via an electric dipole (indicated on the *background*; in this case, pulse intervals were 50 ms, or a pulse rate of 20 Hz). *A* Starting from its hiding tube, the fish typically approaches the Petri dish in one smooth and quick swimming bout within as little as 2s for the 30 cm distance (time measured from stimulation onset). *B* The fish extracting its reward, a *Chironomus* larva, from the feeding tube. The other tube serves for injecting a few air bubbles as a punishment should fish try to get a reward on the presentation of the "wrong" stimulus, S−. *C* A trained fish responding to the S− (in this case, a pulse train of 38 ms intervals). Instead of approaching the feeder, the fish attacks the stimulus dipole. Redrawn from photographs of video frames; fish No. 2

The experimental tank (75 × 40 × 42 cm high) was provided with a hiding tube at the rear center and an electric dipole model near one front corner (as in Fig. 4.32, except that there was only one dipole). Both hiding tube and dipole model were oriented in parallel to the small sides of the aquarium. Near the dipole model a small glass Petri dish (diameter, 5 cm; height, 1.3 cm) with two vertically oriented, fine glass tubes served as a feeder: when fish responded correctly, a single *Chironomus* larva was pushed to the opening of one of the glass tubes by hydrostatic pressure, using a manually operated 20–ml syringe connected to the glass tube by plastic tubing (the opening of the glass tube was clear of the bottom of the Petri dish; inner tube diameter, 2 mm; see Fig. 4.41). Fish responding incorrectly were "punished" by a few air bubbles delivered by an identical second glass tube/syringe arrangement in parallel to the first.

Training and test sessions were performed during light hours only (L:D, 12:12 h) when fish would return to their hiding tubes after obtaining a reward (or punishment). Temperature was $27\pm1°C$; water conductivity, 110 ± 10 $\mu S \cdot cm^{-1}$; the water was aerated and filtered.

Table 4.3. Sex or age correlated characters of two adult fish trained to discriminate constant rates of weak electric pulses (*Pollimyrus isidori*, 7–8 cm length)

P. isidori	Latency PLR/PLA	EOD waveform	Anal fin shape	Anal fin reflex
Fish No. 1	PLR	female	fem/juv	+
Fish No. 2	PLR	male	male	−

The intensity of the stimulus field was $1.2\ \text{mV}\cdot\text{cm}^{-1}$ at the resting position of the fish (measured with fish and hiding tube removed). This matched the intensity of an EOD of a *P. isidori* at the place of the electric dipole (separation of the vertically oriented carbon rods of 3 mm diameter, 2.6 cm). Monopolar square wave pulses of 200 µs were digitally synthesized by a microprocessor-controlled D/A-converter with memory (Kramer and Weymann 1987). This device was controlled by a BASIC-program run on a small desktop computer. Interpulse intervals varied below ± 10 µs (or 0.01% at 100 ms).

Both fish used were males, as shown by their type of latency-related response (Preferred Latency Response, PLR; see Fig. 4.11), although fish No. 1 had an EOD waveform more typical of females (see Fig. 3.4), and fish No. 2 did not show an anal fin reflex [Table 4.3; the anal fin reflex is believed to indicate the male sex in *P. isidori* (Kirschbaum 1987); but cannot be observed in some males (Bratton and Kramer 1988)].

Stimulation was started only when fish were resting in their hiding tube and was stopped after 90 s at the latest, (for example, when fish did not respond). Training success was measured as the latency from the onset of S+ to the fish's arriving at the feeder, where it assumed a characteristic posture while extracting the food reward from the opening of the reward tube (Fig. 4.41B). At that moment the stimulus was stopped. Fish would accept 20–40 larvae per day by means of this training procedure. The inter-trial interval was a minimum of 30 s (or 3 min plus 30 s) and was determined by throwing dice, with 30 s added for each point. However, the interval was longer when fish had not returned to their shelter.

When fish reliably associated a certain stimulus pulse rate (S+) with a food reward, another pulse rate was introduced as the S−. Both pulse rates were used equally often and in random sequence. At the beginning of discrimination conditioning, fish would swim to the feeder also on stimulation with S−, and were punished by a few air bubbles delivered by the second tube arrangement. After learning, fish showed typical responses also to the S−. Fish No. 1 either would not leave its shelter at the presentation of S−, or, when it did, it would stop short of the Petri dish within a distance of a few centimeters. Fish No. 2, however, would attack the dipole electrodes rather than approach the feeder in what looked like "redirected aggression" (Fig. 4.41C). A typical record of the initial phase of discrimination learning is shown in Fig. 4.42.

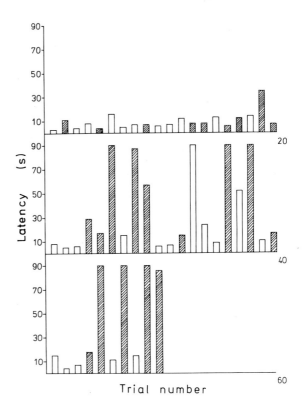

Fig. 4.42. Discrimination learning of *Pollimyrus isidori* No. 1. *Open bars* Rewarded stimulus S+ (here, 10 pulses per s); *hatched bars* S− (here, 25 pps). *Ordinate* "Latency" [s] from the onset of stimulation to the fish's touching the feeder; *abscissa* trial number. After the fish had associated the 10-pps stimulation with a food reward, it initially also took the S− of 25 pps for a positive announcement (the fish generalized). When the fish consistently received a mild punishment (a few air bubbles) instead of a reward at the feeder, latencies to the S− gradually became longer while those for the S+ tended to remain short. When the fish did not respond stimulation was stopped after 90 s

When fish discriminated both pulse rates (as shown by consistently shorter latencies to the S+ as compared with the S−) "test" experiments were introduced. Every fourth stimulation was a test experiment, with random selection of S+ or S−. In test experiments, neither the S+ nor the S− were reinforced (no reward, no punishment). Statistical tests of the fishes' discrimination performance, as presented here, are based on test experiments, exclusively (Mann-Whitney U-test; see the example of Table 4.4).

In the first experiment the S+ was set to 100 ms, close to an EOD interval observed most often in adult *P. isidori* (for example, resting or courting; Figs. 3.24, 4.15, 4.16), and also in young fry (Westby and Kirschbaum 1977; Bratton and Kramer 1989).

Both fish discriminated an S− of 40 ms intervals (a pulse rate of 25.0 Hz) from an S+ of 100 ms intervals (10.0 Hz, Fig. 4.43; Table 4.5). In subsequent experiments, the S− was progressively made more similar to the S+. Fish discriminated up to about 80 ms intervals (fish No. 2: 81 ms) from the S+ of 100 ms intervals, and seemed incapable of performing better. However, when fish were retrained to recognize a *longer* inter-pulse interval (lower pulse rate) from the S+ remaining constant, fish No. 1 still discriminated an S− of 103 ms from the S+ of 100 ms intervals (a 3% difference), while fish No. 2 seemingly failed to discriminate an S− of 133 ms from an S+ of 100 ms intervals. A low performance such as this most likely indicates a (temporary) indisposition of the fish to adapt its behavior to the rules of a game defined by

Table 4.4. Discrimination of constant rates of electric pulses in *Pollimyrus isidori*[a]

Pulse intervals (ms)		n	Mean latency (s)	Standard error (s)	P
S+	50.0	6	4.0	1.22	0.02
S−	42.0	4	69.3	20.75	
S+	50.0	5	2.6	0.22	0.01
S−	44.0	4	90.0	0.00	
S+	50.0	4	3.0	0.70	0.005
S−	46.0	7	72.6	11.30	
S+	50.0	7	3.4	0.42	0.05
S−	47.0	4	47.0	21.70	
S+	50.0	3	3.0	0.00	0.05
S−	48.0	3	90.0	0.00	
S+	50.0	5	20.8	17.31	0.025
S−	48.5	15	44.9	9.76	
S+	50.0	4	4.3	0.85	0.05
S−	49.0	3	42.3	23.96	
S+	50.0	3	19.3	12.2	0.05
S−	53.0	3	69.7	20.5	

[a] Sequence of experiments, from top to bottom (fish No. 2; *bottom pair of lines* fish No. 1, best performance). Example of an S+ of 50 ms. S+ Rewarded stimulus (constant inter-pulse interval in ms, presented for up to 90 s), S− unrewarded stimulus associated with mild punishment should fish try to receive a reward. *n* Number of "test" experiments (no reward, no punishment). *Latency* the time from stimulus onset to the fish's touching the feeder tube; *P* level of significance of the difference between one set of paired results (S+, S−; Mann-Whitney U-test; (Full documentation: Heinrich 1987).

Tabelle 4.5. Discrimination threshold for constant pulse rates (electric pulses for *Pollimyrus isidori*, sound clicks for the two human subjects), testing for the perception of "absolute pitch"[a]

S+ (Intervals, ms)	50	100	220
(Pulse rate, Hz)	20	10	≈4.5
Human subject No. 1	−2%	+3%	+1.6%
Human subject No. 2	−8%	+6%	+3.2%
Pollimyrus isidori	−2%	+3%	+5%

[a] S+ Pulse rate that was rewarded in the case of the fish (expressed as Hz; or constant inter-pulse interval, ms). The table gives the percent difference (based on ms) between an S+ and the S− (unrewarded pulse rate) which was closest to the S+, and still discriminated. For example, at an S+ of 100 ms pulse intervals, *P. isidori* recognized longer intervals down to 103 ms as the S−, or a +3% difference from the S+.

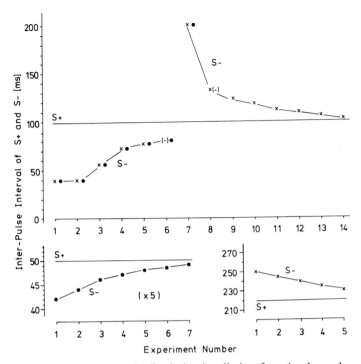

Fig. 4.43. Summary of *Pollimyrus isidori's* discrimination limina for stimulus pulse trains of constant rate, as tested by a conditioned discrimination paradigm (see Figs. 4.41, 4.42). *Ordinate* Inter-pulse intervals (ms); *abscissa* experiment number. $S+$ Pulse rate associated with a food reward, $S-$ pulse rate associated with mild punishment should fish try to get a reward. *Xs* fish No. 1, *Os* fish No. 2. When fish discriminated, the $S+$ was kept constant while the $S-$ was made more similar to the $S+$ in the next experiment (and so on). *Top Half* At first, both fish seemed unable of discriminating interpulse intervals of a less than about 20% difference (because they did not discriminate intervals greater than about 80 ms for the $S-$, from an $S+$ of 100 ms intervals (experiment nos. 1–6). However, when the pulse intervals for the $S-$ were *longer* than those for the $S+$, fish No. 1 still reliably (and statistically significantly) discriminated an $S-$ of 103 ms intervals from an $S+$ of 100 ms intervals (or a 3% difference; experiment no. 14). *Lower Half* Fish's No. 2 best performance was the discrimination of an $S-$ of 49 ms from an $S+$ of 50 ms intervals (a 2% difference; experiment no. 7, *lower left*; *ordinate* magnified x 5), and fish No. 1 discriminated an $S+$ of 220 ms from an $S-$ of 231 ms intervals (a 5% difference; experiment no. 5, *lower right*). In spite of considerable effort at each one of these three $S+$ intervals, no finer discrimination was achieved

the human observer, or a lack of skill of the experimenter so that the fish cannot associate the $S+$ with reward nor the $S-$ with punishment, or both. A year later, this fish No. 2 discriminated an $S-$ of 95 ms from an $S+$ of 100 ms ($P<0.05$).

The next experiment with fish No. 1 was devised to test whether its discrimination performance (of an $S-$ of 103 ms from an $S+$ of 100 ms) could be improved still further. We started with an $S-$ of 103.0 ms and planned to gradually glide down to 102 ms, but the fish no longer discriminated. It continued, however, to discriminate an $S-$ of 104 ms from the $S+$ of 100 ms.

Fish No. 2 excelled in a later discrimination task: it still discriminated an S+ of 50 ms intervals from an S− of 49 ms (a 2% difference). This result could not be improved, nor could it be demonstrated on each day. When the fish failed to discriminate 49 from 50 ms it still succeeded in doing so when the S− was 48 ms. The central EOD interval of a swimming *P. isidori* is close to 50 ms (see Fig. 3.24). At an S+ of 50 ms, the best performance of fish No. 1 was the discrimination of a 6% difference, an S− of 53 ms ($P<0.05$). However, this experiment was performed a year later than the others when the fish were, apparently, no longer in prime condition.

A third series of experiments used a pulse interval of 220 ms as the S+. That interval is close to the low-frequency EOD interval mode of resting *P. isidori* (see Fig. 3.24). Starting from an S− of 250 ms intervals, fish No. 1 still discriminated 231 ms intervals from the S+ (220 ms intervals). This is an only 5% difference. It was not possible to improve, nor to repeat this performance; however, the fish continued to discriminate an S− of 232 ms from the S+ of 220 ms.

Two human subjects in their early twenties were tested for their discrimination limina of constant pulse rates under conditions resembling those for *P. isidori* as closely as possible. Of course, the output device (the electric dipole) had to be replaced by a loudspeaker, and the food reward by informing the subject immediately whether his or her judgment had been correct.

Only at the lowest pulse rate, an S+ of 220 ms intervals (or about 4.5 Hz), did both human subjects perform better than *P. isidori* in our tests (Table 4.5). At pulse rates of 10 and 20 Hz, however, fish discriminated as well as humans in the case of a trained musician (subject No. 1), and considerably better than the other human subject (8 and 6% at 50 and 100 ms intervals, respectively).

P. isidori's thresholds for pulse rate differences are comparable with or even better than the lowest frequency difference thresholds known for any animal in the realm of audition and of perception of water surface waves, another lateral line sensory function, at least when the S+ and the S− are presented separately at substantial intervals of time (as in the present experiments). This experimental procedure tests for the capacity of "absolute pitch", only rarely found among professional musicians. For example, the clawed frog's *Xenopus* discrimination limen for water surface waves of different frequencies is ±3% at 15 Hz (Elepfandt 1985). Discrimination limina for hearing usually improve considerably when the two stimuli are contrasted directly, for example, by presenting them in alternation (Dijkgraaf and Verheijen 1950).

Not only advanced vertebrates like mammals, but also teleost fish, lower vertebrates lacking a cochlea as they do, were shown to be able to discriminate sound of different frequencies (Stetter 1929; Wohlfahrt 1939, both for ostariophysine fish; Stipetić 1939 for a mormyrid; reviews Fay 1974; Popper et al. 1988). Dijkgraaf and Verheijen (1950) reported that the minnow (*Phoxinus*) was capable of discriminating tones with a 3% difference (or a quarter tone on the musical scale). Comparable values, this time with stimulus intensities adequately controlled, were found in the goldfish, *Carassius* (Jacobs and Tavolga 1968; Fay 1970). Ostariophysine fish like the goldfish have a much better hearing sensitivity and a wider frequency range than most other teleosts; frequency difference thresholds of the few non-ostariophysines tested range from 10 to 100% (reviews Tavolga 1971; Popper and Fay 1973). Stipetić's (1939) value of 19% for the mormyrid *Gnathonemus* (probably *Marcusenius macrolepidotus*) should be regarded as highly preliminary, like most other threshold reports for fish of the first half of this century (Jacobs and Tavolga 1968; Fay 1970). The lowest frequency difference threshold for humans is 0.3% at the optimal frequency range around 1000 Hz (reviews Roederer 1975, Zwicker 1982).

More comparable to the present experiments are investigations in which animals were trained to detect changes in the repetition rate of sound clicks or noise bursts (reviewed in Fay and Passow 1982). The relationship of just noticeable differences (JNDs) in stimulus period or duration (ms) over period or duration (ms) approximately conforms to Weber's Law in man, parakeet, and porpoise (a power function with an exponent equal to 1; see Fay and Passow's Fig. 5), while the goldfish shows a steep decline in temporal acuity toward long durations between sound clicks (a power function with slope 2). For example, at 50 ms repetition period (the longest investigated), the goldfish's just discriminable change in period is about 23 ms in the best case ("type 2 gated noise", Fay and Passow's Fig. 4). This is a very high threshold of 46%, while it is only 3.2% at a period of 5 ms (or 200 Hz repetition rate). The latter value was nearly equal to the discrimination threshold found for a 200-Hz pure tone, and is therefore not comparable to the present experiments. Fay (1982) has shown that in many auditory neurons in the goldfish, inter-spike interval distributions obtained in response to 200-Hz periodic bursts are indistinguishable from those obtained for 200-Hz pure tones.

The lowest thresholds for long repetition periods of tone bursts (periods below 1 s) are those from man and porpoise. Up to at least 300 ms (man) and 600 ms (porpoise), the discrimination threshold remains constant at around 6 to 7% (review Fay and Passow 1982, their Fig. 5).

This shows that the discrimination acuity of *P. isidori* for intervals of electric pulses is exceptionally good, even though experiments contrasting different electrical pulse rates *directly*, as in the other more modern examples given above (excepting the clawed frog), have not been tried. *P. isidori*, and probably other mormyrids as well, appear to be specialists in the discrimination of relatively long inter-EOD-intervals. It would be interesting to know whether this sensory acuity is limited to the electrical sensory modality in these fish.

The exceedingly high performance of *P. isidori* in discriminating differences of time intervals may not be totally surprising, given the huge cerebellum of mormyrids, unrivalled among vertebrates (see Sect. 2.1.4). Braitenberg (1977) views the vertebrate cerebellum not only as a timing organ for precise and rapid motor coordination but also as a clock (stopwatch) that can measure time intervals. According to his considerations, largely based on anatomy, the shortest interval of time the vertebrate cerebellar cortex can resolve should be about 0.1 ms.

4.2.2 Gymnotiformes

4.2.2.1 Electrical Stimulation and Playback of EOD Patterns in Pulse Species

Attack-Eliciting Stimuli in a Pulse Species. When one electric eel discharged at moderate to high frequency it attracted other eels that also increased their EOD rates (Bullock 1969). These observations, which involve the eel's strong discharge, may have been similar to those of Cox (1938), who observed mutual attraction among attacking and feeding eels.

Black-Cleworth (1970) and Westby (1975a,b) have investigated the social behavior of another pulse species, *Gymnotus carapo*, in more detail (see Sect. 4.1.2.1). Both authors also studied the stimuli evoking attack responses in this species experimentally. Black-Cleworth investigated a wide scope of stimuli and concepts, while Westby (1974) focused on the effect of electrical pulse patterns by playback (see below).

In Black-Cleworth's study, *G. carapo* attacked conspecifics much more frequently than it attacked non-electric fishes (*Lepomis cyanellus*, the green sunfish, Centrarchidae; and *Xenomystus nigri*, the African knifefish or "false featherfin" of rather gymnotiform appearance, Notopteridae; although only distantly related, this fish was recently discovered to be electroreceptive; review Braford 1986; Chap. 2). "Heterospecific electric fishes were generally intermediate in eliciting attack, with those whose discharges were most similar to *Gymnotus* being attacked more frequently." (The fishes attacked more frequently were *Hypopomus beebei* and especially *Steatogenys elegans*; their discharges were most similar to those of *G. carapo* in both pulse shape and rate among the fishes used.) "Electrodes emitting pulses were attacked more frequently than those without pulses and such visual stimuli as mirrors or fish within glass jars. Anaesthetized conspecifics were attacked more frequently when they were emitting pulses than when they were silent."

These experiments indicated that electrical discharges, much more than visual stimuli, were important in eliciting attack, and that some species specificity was involved.

"The differences in attacks upon (fish) models of different materials and sizes were relatively small. ... The addition of electrical pulses greatly increased attacks upon all models, ..." Black-Cleworth concluded that "active production of electric current is of primary importance in eliciting attack and that this current must change in time; DC was ineffective compared to pulses."

"Short duration, rectangular pulses elicited attacks at lowest intensities at frequencies of 100 to 500 Hz; thresholds rose relatively sharply at higher frequencies and were more constant at lower frequencies. Attack thresholds for sinusoidal frequencies (that is, sine wave stimuli) were lowest at 500 to 1000 Hz and rose rather sharply at both lower and higher frequencies. Constant frequency pulses elicited attack at low threshold intensities close to the fish, and thresholds increased sharply with distance" (Black-Cleworth 1970).

Black-Cleworth also studied experimentally the conditions evoking EOD rate modulations ("signals") by stimulation with models, with and without added electrical pulse patterns. Part of the patterns used were artificially generated, part were tape-recorded natural patterns [such as SIDs (sharp increases in frequency followed by decreases to original level) and breaks].

SIDs and the motor act of serpentining appeared to be evoked by the same stimuli (of a model fish) that elicit attack in social context. "Discharge cessation was common only when the stimulus object emitted electrical pulses; in this situation, breaks, but not arrests, were commonly associated with attacks."

"Breaks were often elicited by lower intensity electrical stimuli than by either attacks or arrests, and arrests tended to have higher thresholds than attacks." The resident fish seemed to be "using an intensity measure related to voltage to estimate the size of the approaching 'electrical enemy'."

"Fish exposed to situations where electrodes in a hiding place emitted various signals tended to approach and retreat more often when SIDs and breaks were given by the electrodes. They entered and stayed in the hiding place more often and longer when constant frequency pulses were on the electrodes. SIDs, breaks, and attacks by the fish occurred when pulses and/or signals were emitted by the electrodes, but were rare when no pulses were on the electrodes" (Black-Cleworth 1970).

Westby (1974) used four artificially generated, simulated discharge patterns in his playback experiments with *G. carapo*. The artificial fish bore no physical resemblance to a conspecific; however, it resembled it electrically (electrode separation, pulse duration, and intensity). Westby's stimulating device was a dipole floating electrically, while Black-Cleworth used monopoles referenced to a distant ground electrode.

Aggressive behavior towards the model was only shown when it was electrically active, and was always directed to the same end of the dipole, unless polarity of the diphasic pulses (roughly resembling *G. carapo*'s EODs) was reversed. The frequency distributions of motor acts and interaction with the model revealed great similarity to the results obtained with pairs of fish (Westby 1975a,b).

Both studies show the overriding importance of electrical pulses for eliciting attack and other social behaviors in *G. carapo*.

Display-Evoking Stimuli in a Pulse-Species. Female *Hypopomus occidentalis* give "decrement bursts" both in courtship and aggressive interactions (brief accelerations of their 50–110 pulses/s EOD rate, involving 3–10 EODs; Hagedorn 1988). Females have shorter EODs than males (see Sect. 3.1.2.1). In stimulation experiments using single-cycle, symmetrical sinusoids, female *H. occidentalis* gave more decrement bursts when the duration of the sinusoidal stimulus pulses resembled that of the relatively long male EODs compared to shorter stimulus pulses of female-typical duration (Shumway and Zelick 1988). These authors consider frequency discrimination based on spectral tuning of elecroreceptors the most likely neuronal mechanism that underlies this differential sensitivity.

In these experiments, the stimulus pulse rate was varied according to the fish studied; the effect of pulse rate on the outcome of the experiment was, apparently, not studied systematically.

Jamming Avoidance and Phase Sensitivity in Pulse Gymnotiforms. By discharging its electric organ, a weakly electric fish tests its environment. A change of feedback very often evokes a change of EOD rate (Fig. 3.32). Feedback change may have two causes: (1) change of resistivity of the medium, as when an object approaches; (2) coincidence of the fish's own EOD with that of a near neighbor. Mormyrids seem to be bothered very little by the second possibility, for their EOD rates are generally low, EOD pulse durations short (except for a few species), and successive EOD intervals variable. Therefore, the probability of an EOD coincidence in a pair of mormyrids is very low, indeed (partial or total coincidence of EOD main phases, up to three occurrences in 1000 pairs of pulses for *Pollimyrus isidori*; see Lücker and Kramer 1981); the probability of a whole series of coincidences is so low we need not discuss further. Mormyrids have a "built-in" protection from coincidences of EODs occurring in series, without doing anything special. In addition, they appear protected from the confusion caused by other fishes' EODs by a specialized sensory mechanism relying on an efference copy of their own discharge (Zipser and Bennett 1976b; Sect. 2.1.4). As if this were not enough, at least some mormyrids display phase-locking (Preferred Latency Responses, PLRs, or Preferred Latency Avoidances, PLAs) to other fishes' pulse EODs, which probably aids in coping with the disruptive sensory input arising from other fishes' EODs (Sects. 2.1.3, 4.1.1.1).

Not so, however, in gymnotiform pulse species. Compared with mormyrids they appear stricken by all possible disadvantages: (1) their pulse EODs tend to be longer, (2) EOD rates are often higher, compared to those of mormyrids; that is, the duty cycle, pulse duration/pulse interval, is greater and so is the probability of pulse coincidences, and (3) EOD rates of gymnotiform pulse species are usually much more stable than those of mormyrids; that is, the time the next EOD will occur can be predicted within narrow limits from the preceding two EODs in gymnotiforms. The EOD rates of some pulse gymnotiforms are even as constant as those of wave fish (on an interval-to-interval basis).

A pair of pulse gymnotiforms will run into trouble whenever fish are sufficiently close to each other physically and in frequency, especially fish of high and constant frequency like *Rhamphichthys* and *Hypopygus*. If the frequency difference, ΔF, is low, the EODs of one fish will drift slowly and regularly through all phases of the other fish's EOD cycle, including a series of coincidences of the fishes' EODs (when they are "in phase"). The smaller ΔF, the longer the sequence of partial, full, and again partial overlap of pulse waveforms will be before the EODs are out of phase again. Depending on ΔF, a sequence of EOD coincidences will be regularly repeated at intervals. Clearly, this situation must be seriously disturbing the sensory feedback from a fish's own EOD, making it temporarily "blind" for any object-induced changes in feedback.

Fish promptly take action, however, although species differences are apparent. Especially interesting are experiments using EOD-triggered stimuli with controlled phase relationships to the EOD cycle. The first such report was by Larimer and MacDonald (1968). In a low-frequency *Hypopomus* with a 4-ms EOD repeated at 3–5 Hz, transient frequency increases were evoked by such "in-phase" stimulus pulses. Thresholds were lowest when stimulus pulses coincided with EODs, and up to 25 times higher at delays in-between EODs. A similar response was evoked by a change of the medium resistance, which also altered EOD amplitude. Very similar results were obtained in *Hypopomus occidentalis*, discharging mostly at much higher and more regular rates (Bullock 1969). He explained the fish's sensitivity to stimuli coinciding with its EOD by peripheral sensory mechanisms (see Sect. 2.1.3).

Gymnotus carapo, when stimulated in phase with, or during, its EOD pulse, also proved ten times more sensitive than when stimulated in-between EODs, and responded by transient EOD rate increases. The explanation of this high sensitivity was similar to that mentioned above: the sigmoid stimulus/response curve common to most receptors means keen sensitivity to minute intensity differences of strong stimuli (such as the fish's own EOD) in spite of a low absolute sensitivity (MacDonald and Larimer 1970).

The behavior of *Hypopygus*, a pulse fish with a quite constant pulse rate, resembled the jamming avoidance response in a wave fish, *Eigenmannia*: when stimulated with a train of free-running pulses at a small ΔF, it changed its frequency such that ΔF increased in both directions (Heiligenberg 1974; however, only a single specimen could be studied). The time courses for increase and decrease were not similar, however. During an increase, the fish briefly accelerated whenever it passed pulses in a slower train, thus avoiding long series of pulse coincidences. The frequency over time record had phasic spikes superimposed on a tonic increase, while the frequency decrease was tonic and smooth. *Hypopygus* proved as sensitive to stimuli coinciding with its EOD as to those just preceding it within 2 ms; hence, the explanation of the fish's high sensitivity by the sigmoid receptor curve (see above) is insufficient in *Hypopygus*.

Westby (1975c) found four response types to EOD-triggered stimuli in *Gymnotus carapo*: "transient frequency increase", "transient frequency decrease" (both followed by a return to resting EOD frequency), "off", and "phase oscillation" to the effect of EOD coincidence avoidance (Westby referred to the last behavior as "jamming avoidance", although other authors describe by this term an active and usually tonic increase of the absolute value of ΔF by the fish). Offs occurred at all latencies but the other response types appeared with a probability that was a function of the delay. Transient frequency decreases were predominantly seen to stimuli just before, during, and just after the fish's own EOD (in contrast to the observations of MacDonald and Larimer 1970, see above), while transient frequency increase was the commonest response type to stimuli in-between EODs. A preferred latency of 13 ms in pairs of fish discharging in synchrony ($\Delta F = 0$ Hz) was also found, which may be a kind of "phase coupling" (see Langner and Scheich 1978; Sect.

4.2.2.2; Gottschalk and Scheich 1979), just like the above-mentioned phase oscillations (see also Sect. 4.1.2.1; Westby 1979).

Gymnotus' threshold is at maximum immediately after an EOD and remains high for the first half of the EOD interval (about $1 \text{ mV} \cdot \text{cm}^{-1}$), but drops rapidly at a delay of 12.5 ms to the low level found before and during an EOD ($10 \text{ μV} \cdot \text{cm}^{-1}$). Westby's results not only show that the fish's sensitivity may be just as high outside the fish's EOD as within (which is incompatible with an explanation relying on the sigmoid receptor curve just as in *Hypopygus*; see above), but also that there are excitatory and inhibitory phase regions within a discharge cycle. The phase oscillations suggest a control mechanism, especially in view of the results in *Rhamphichthys* (Scheich et al. 1977; Gottschalk and Scheich 1979):

Scheich et al. (1977) found a jamming avoidance response very similar to that of *Hypopygus* in a *Rhamphichthys* (reportedly *rostratus*), another fish with an extremely constant discharge frequency. (Both species' EOD is much more periodic than, for example, that of *Gymnotus*.) When all stimulus pulses coincided with the EOD, phasic-tonic EOD frequency increases were observed (Fig. 4.44C). Unlike in *Hypopomus*, phase-locked stimuli before and after this sensitive phase elicited a frequency decrease (Fig. 4.44D). There were sharp boundaries between phase ranges where stimulus pulses had excitatory or inhibitory effects on the pacemaker (Fig. 4.44A). During stimulation with a free-running pulse rate of small ΔF, there is, therefore, a sequence of excitatory and inhibitory effects, as the stimulus pulses drift through all phases. This sequence occurs in opposite order for $+$ and $-\Delta F$.

Scheich et al. (1977) showed that besides direct excitatory and inhibitory effects on the pacemaker, as manifested by phase-locked stimulation, there exists a sensitivity for the *direction of phase shifts* relative to the EOD, as seen in ΔF stimulation. This directional sensitivity was only observed when a boundary between excitatory and inhibitory sensitive phase ranges was crossed. Therefore, ΔF stimulation with gated stimuli which remained entirely within the boundaries of a particular phase range (representing an excerpt of a ΔF stimulus) gave only frequency decrease in the inhibitory phase range (for example, Fig. 4.44Bj), and frequency increase in the excitatory phase range (Fig. 4.44Bh), with both $+$ and $-\Delta F$ stimuli (unlike with free-running ΔF stimulation). When the delay of these gated stimuli was changed such that they crossed a phase boundary, a "correct" response was always obtained (that is, frequency decrease to a higher frequency stimulus (Fig. 4.44Bk), and frequency increase to a lower frequency stimulus).

Thus, the JAR in *Rhamphichthys* seems to require a key stimulus fundamentally different from that in wave fish. The response appears to rely on a sensitivity for the direction of phase shifts of stimuli relative to the own EOD, rather than on a sensitivity for beats as in the JAR of the wave fish *Eigenmannia* (Scheich et al. 1977).

A similar idea was set out in Heiligenberg (1977) and Heiligenberg et al. (1978a) for other periodically discharging, gymnotiform pulse fishes. For phase coupling behavior (Sect. 4.2.2.2) in *Rhamphichthys* see Gottschalk and Scheich (1979).

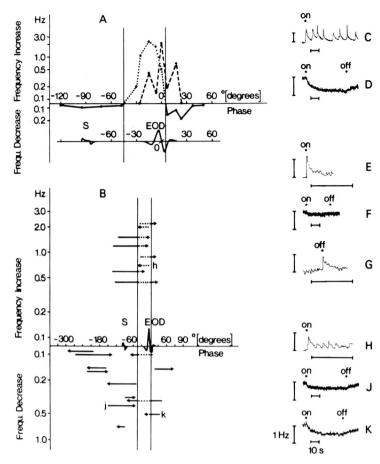

Fig. 4.44. Responses of *Rhamphichthys* to phase-locked and phase-shifting (ΔF) stimulus pulses, measured as frequency changes of its electric organ discharges (EODs). *C, D* Jamming avoidance responses to + and –ΔF stimulation, respectively. *A* The increase and decrease of the fish's frequency above and below the resting frequency is plotted as a function of the phase at which a phase-locked stimulus pulse occurs. The reference point on the abscissa is the center zero crossing of the EOD (zero phase). *Solid line* On-responses in the two inhibitory phase ranges; *points* the phasic on-responses to stimulus pulses within the excitatory phase range; *broken line* the phasic off-response in certain parts of excitatory and inhibitory phase ranges. For the on-response, stimulus phase ranges which induce frequency increases and decreases, are separated by *vertical lines*. The fish's EOD with a stimulus pulse (*S*) is shown below. Responses to phase-locked pulses are shown in *E* and *G* (on- and off-responses in the excitatory phase range) and *F* (on-response in an inhibitory phase range). *B* The increase and decrease of the fish's frequency to gated, phase-shifting stimuli. *Arrows* mark the range and direction of the phase shift of the stimuli on the abscissa. The vertical position of the arrows indicates the magnitude of the EOD frequency shift. Arrows within the excitatory phase range are *dotted*. Arrows h,j,k correspond to the frequency records *H, J, K* to the *right*. *Vertical scales* in *C–K* indicate 1 Hz frequency change of the EOD, *horizontal bars* indicate 10 s. The resting frequency of the fish is represented by the initial value of each record in *C–K* (40–70/s in different fish; Scheich et al. 1977)

4.2.2.2 Electrical Stimulation and Playback of EODs in Wave Species

Behavioral Responses. Mature male *Sternopygus* discharge at lower frequencies than females (see Sect. 4.1.2.2). Hopkins (1974b) observed an increase in number of EOD frequency modulations (such as rises and interruptions) when he stimulated two males showing site attachment in their natural environment, with sine waves through an electrode dipole, at frequencies between 80–250 Hz (distance from fish: 1 m). Responses were inhibited within the male range (55–80 Hz), and at exactly twice the male's frequency. More information about the exact nature of the responses would be desirable.

Hopkins (1974a) also studied responses of *Eigenmannia virescens* towards a dipole model of an electric fish in aquaria (model gymnotid-shaped, constructed from transparent plexiglass). The model was positioned 5 cm above the aquarium bottom. A variety of behaviors towards the electrically active model were seen: darts, butts, and EOD interruptions, but no "rises" (see Sect. 1.2.2). "During the non-breeding season, *Eigenmannia* males and females ($n=2$ males, 4 females) responded to playback of tape-recordings of *Eigenmannia*, by giving Attacks (...) and Discharge Interruptions. The recordings of other sympatric species of gymnotids were significantly less effective in eliciting responses." (These species were *Apteronotus albifrons*, *Sternarchorhamphus macrostomus*, *Sternopygus macrurus*, and the pulse fish *Gymnorhamphichthys hypostomus*, in decreasing order of effectiveness.)

"Sinusoidal electrical stimuli elicited Attacks and Interruptions from five *Eigenmannia* during the non-breeding season when the frequencies were in the range of 200 to 700 Hz. (...) During the breeding season, the situation changed somewhat, in that males responded to playbacks of recorded signals of *Eigenmannia* by giving large numbers of Interruptions (...). The level of attack by males was also increased during the breeding season. Male *Eigenmannia* also gave high levels of Interruptions during the breeding season in response to sinusoidal stimuli with frequencies within the species range." Hopkins (1974a) concluded that "males appear to give Interruptions as a type of courtship display."

This is in agreement with Hagedorn and Heiligenberg (1985), who observed the spawning behavior of *E. virescens* in small aquarium groups. Playback through electrodes of a recording of a courting male's EOD, characterized by frequent interruptions ("chirps" in Hagedorn and Heiligenberg's study), could induce gravid females to spawn. Males did not chirp in response to constant frequencies from sine-wave generators, but did so as soon as the constant frequencies were modulated from 1 to 10 Hz, in the manner of a "long rise" (a display given by females). Fish (especially large males) may, however, display chirps and short rises in response to stimuli of constant frequencies close to their own EOD frequency (Kramer 1987).

Jamming Avoidance Responses and Phase Coupling. Watanabe and Takeda (1963) discovered a "peculiar response" in the wave fish *Eigenmannia*: when stimulated with a weak electric signal (greater than $3 \, \mu V \cdot cm^{-1}$) close to its own EOD frequency, *Eigenmannia* changed its frequency "as if to escape from

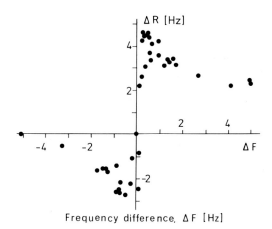

Fig. 4.45. The frequency difference vs response curve of *Eigenmannia* ($n=1$) as observed by Watanabe and Takeda (1963); redrawn from their Fig. 4). The response is measured as the frequency change, $\Delta R = F_{Response} - F_{Rest}$ (Hz). The frequency difference between the fish's resting EOD and the stimulus is $\Delta F = F_{Fish} - F_{Stim}$ (Hz). With the stimulus frequency higher or lower than the fish's resting frequency, the fish lowered or raised its frequency, respectively, increasing the frequency difference. No response was observed at $\Delta F = 0$ Hz. Each point is one single response (After Watanabe and Takeda 1963)

the applied frequency". With a stimulus freqency higher (or lower) than the EOD frequency, the fish lowered (or raised) its EOD frequency. The effectiveness of the stimulus depended on the frequency difference, ΔF: the smaller the difference the more effective the stimulus. The response only failed to occur "when ΔF was very close to zero" (Fig. 4.45; confirmed by Larimer and MacDonald 1968). Watanabe and Takeda (1963) assumed the function of *Eigenmannia's* response to frequencies close to its own was to enable the fish "to distinguish between its own signal and those of its neighbors" for better object detection.

The response was investigated in greater detail by Bullock et al. (1972a,b) and Scheich (1977a), who called it the "jamming avoidance response" (JAR; Fig. 4.46). The optimal ΔF eliciting strongest responses was about ± 3 Hz when using a frequency difference clamp devised to hold ΔF dynamically constant, frustrating the normal escape response from the jamming frequency (Fig. 4.47). This is considerably more than the minimum effective ΔF of below 0.2 Hz (Bullock et al. 1972b), or the optimal ΔF with an unclamped stimulus (0.2–0.5 Hz according to Watanabe and Takeda's plot of Fig. 4.45).

The convergence of electrosensory input on a single output (the medullary pacemaker; Szabo and Enger 1964) governing a quantifiable, socially relevant behavior has stimulated attempts to identify the effective stimulus parameters, the nature of sensory coding, and the correlated central responses (Scheich 1977a,b,c; see review by Heiligenberg 1986). These efforts, along with neuroanatomical studies (see reviews by Scheich and Ebbesson 1983; Heiligenberg 1986), have resulted in neural models of the JAR based on the notion of time

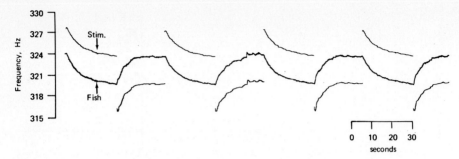

Fig. 4.46. Jamming avoidance response in *Eigenmannia. Stim.*, Sine wave stimulus of $\Delta F = +$ or -4 Hz, held constant for 25 s by a frequency difference clamp, then switching automatically to the opposite sign. Note that the fish's EOD frequency (*Fish*) was chased up and down by the frequency difference clamp (Bullock et al. 1972a)

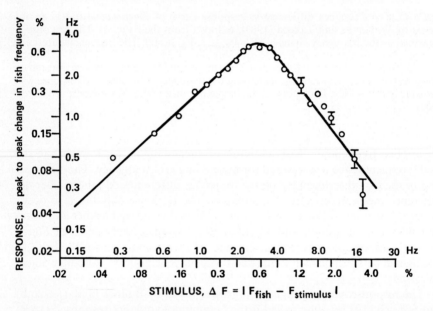

Fig. 4.47. Response as a function of ΔF. The stimulus was applied as alternately plus and minus ΔF of the given (clamped) value, each held for 26 s. *Eigenmannia* of resting frequency 500 Hz (Bullock et al. 1972b)

domain mechanisms of ΔF assessment, as opposed to true frequency analysis (Scheich 1974).

Little work on the JAR has been done with unclamped stimulus frequencies (except Watanabe and Takeda 1963; Larimer and MacDonald 1968; Kramer 1985a, 1987). An unclamped stimulus frequency allows the fish to show its normal escape response, which may or may not be in the direction that would be enforced by the frequency difference clamp. Active frequency following is a type of interaction not yet observed in interacting fish (see Bullock et al. 1972a for evidence on two-animal interactions; Fig. 4.48).

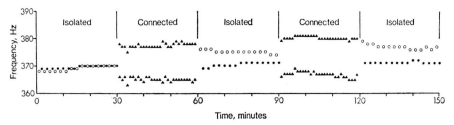

Fig. 4.48. Two *Eigenmannia* selected from a large number for nearly equal electric organ discharge frequency, kept in separate aquaria and measured to ±0.5 Hz, alternately isolated and connected by a pair of wires. Both fish show a jamming avoidance response (Bullock et al. 1972a)

Types of Jamming Avoidance Response. As has already been observed by Watanabe and Takeda (1963), the response, ΔR, to unclamped stimulus frequencies is "very small, if present" when ΔF is more than ±20 Hz. However, the notion that the JAR were free from habituation (for example, Bullock et al. 1972a) could not be confirmed. As experimentally shown, the JAR readily habituates even to very weak stimuli; habituation becomes stronger the shorter the inter-trial interval and the higher the stimulus intensity (Kramer 1985a, 1987; his Fig. 3).

In contrast to earlier studies, some of which have concerned only very few fish, great inter-individual variability in JAR behavior, including responses in the opposite direction to that expected, was found (Kramer 1985a, 1987). Four types of response to unclamped (constant) stimulus frequencies, associated with sex or age, were observed (Figs. 4.49–4.52). Adult males gave no responses or very weak ones (to −ΔF only; Fig. 4.49); adult females gave good responses, frequency decrease, to −ΔF, and no (or only weak) responses to +ΔF (Fig. 4.50). The most effective ΔFs in females (−0.6 to −2 Hz) were considerably greater in absolute terms than found by Watanabe and Takeda (1963) and Larimer and MacDonald (1968; about 0.2–0.5 Hz). The great difference in response strength to negative ΔF values between males and females is not explained simply by lower absolute sensitivity of males resulting from their stronger EOD intensities, as shown by the use of stronger stimuli (Fig. 4.49).

A similar lack of responsiveness to ΔF values of one sign, and responsiveness to values of the opposite sign, is known from two gymnotoid wave fish, *Apteronotus (Sternarchus) leptorhynchus* (Larimer and MacDonald 1968) and *Apteronotus albifrons* (Bullock et al. 1972a,b), but its functional significance has not been discussed (for example, regarding object detection in the presence of jamming stimuli). In these apteronotids, only frequency *increases* in response to +ΔF and no, or only weak, frequency decreases in reponse to −ΔF were elicited.

Sternopygus (of the same family as *Eigenmannia*, Sternopygidae) was reported to lack a JAR by Bullock et al. (1975, p. 117; "a good many individuals" of unspecified size or sex were investigated) and Matsubara and Heiligenberg (1978; three adult, enucleated *S. macrurus* males; frequency

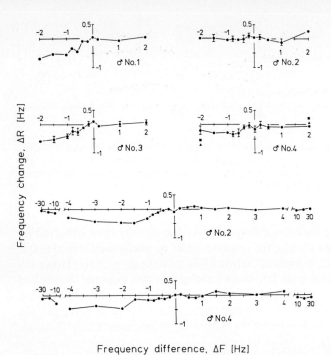

Fig. 4.49. Frequency difference vs response curves of four adult male *Eigenmannia lineata* (*lower two graphs* two males retested 2.5 years later). ΔR and ΔF as defined in Fig. 4.45; stimulus frequency unclamped, that is, constant throughout experiment. Note that the fish were insensitive to +ΔFs (including ΔF=0 Hz); –ΔFs evoked only weak or no responses, which contrasts with the females' behavior (Fig. 4.50). Each point is a mean ± 1 SE ($n=12$). *Standard errors* are either shown or are too small to be drawn. Stimulus intensity, 30 μV · cm^{-1}; except for the *squares* in the plot of male No. 4 (earlier measurement, *upper right*) which are +10 dB, and the *triangles* which are +20 dB (x 10). Inter-trial intervals, at least 20 min (Kramer 1987)

changes of only up to ±0.5 Hz). As neither study provides data, it is not clear whether *Sternopygus* displays only a weak, but statistically significant, response (like *Eigenmannia* males for –ΔF values, Fig. 4.49) or no response at all.

In one group of juvenile *Eigenmannia*, +ΔF stimuli elicited only a small change in frequency (much smaller than noticed by Watanabe and Takeda 1963; and Larimer and MacDonald 1968; Fig. 4.51). Only the other group of juveniles (probably males) gave good responses to any ΔF within the effective range originally described (Fig. 4.52). But even their behavior differs from that described in the earlier studies: these juveniles still *increase* their EOD frequency at –ΔF (down to about –0.5 Hz, and occasionally to –1.5 Hz) when it would be more economical and faster to *decrease* EOD frequency to escape from the jamming stimulus frequency (one additional juvenile male showed a frequency decrease in reponse to ΔF=0 Hz and small +ΔFs).

Also, in contrast to the observations of Watanabe and Takeda (1963) and Larimer and MacDonald (1968), an increase in frequency was stronger than a

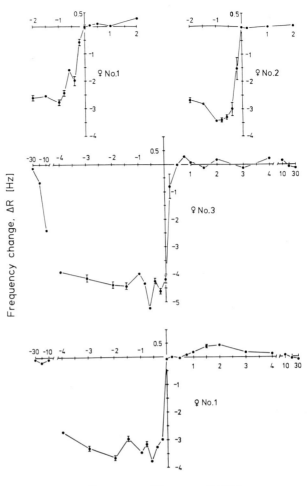

Fig. 4.50. Frequency difference vs response curves in *Eigenmannia lineata*. Like Fig. 4.49 but for three adult females, all gravid with eggs (female No. 1 retested 2.75 years later, *below*). Stimulus frequency unclamped, that is, constant throughout experiment. Note that the females were almost insensitive to $+\Delta$Fs (including $\Delta F = 0$ Hz, except female No. 3); $-\Delta$Fs evoked strong responses in all fish (Kramer 1987)

decrease in these fish. This may be related to the asymmetry of the ΔF curves with respect to the abscissae: in increasing an initial frequency difference of, for example, $\Delta F = -1$ Hz by a response of $\Delta F = -2.5$ Hz (a frequency decrease), a total frequency difference of 3.5 Hz results. That value is only obtained by a much greater increase in frequency of $\Delta R = +3.7$ Hz in response to an initial ΔF of -0.2 Hz (this ΔF value elicited frequency increase, contrary to expectation).

Another puzzling aspect is that at $\Delta F = 0$ Hz, which was ineffective in the other fish (except female No. 3, Fig. 4.50), the juveniles of Fig. 4.52 gave maxi-

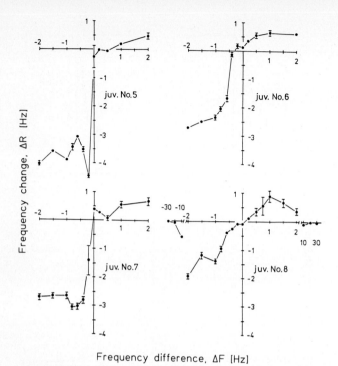

Fig. 4.51. Frequency difference vs response curves in *Eigenmannia lineata*. Like Fig. 4.49 but for four juveniles (or subadults). Stimulus frequency unclamped, that is, constant throughout experiment. These fish gave only weak responses to stimulus frequencies lower than their own ($+\Delta F$), while $-\Delta F$s evoked good responses (Kramer 1987)

mal responses ($+\Delta Rs$). The frequency difference vs response curves given by Watanabe and Takeda (1963) and Larimer and MacDonald (1968) correspond best to those of these juveniles, except that the responses to $\Delta F = 0$ Hz were not observed in any of the earlier studies. Bullock et al. (1972a,b) and Heiligenberg et al. (1978b) also found $\Delta F = 0$ Hz to be ineffective (the number of fish on which this result is based is not clear in either study). One may only speculate about the mechanisms and functions of this divergent behavior ("strategies") in members of one species.

Mechanisms of Divergent Frequency Change Behavior. A stimulus with a specified ΔF, associated with a certain electroreceptor response pattern (see below), does not necessarily lead to one behavioral reponse pattern (JAR) in *Eigenmannia*: this would be expected if the only (or main) function of the JAR was to improve the signal-to-noise ratio for better electrolocation in the presence of stimuli of similar frequency (the "signal" here would be the fish's own EOD). The complex "block diagram of components in the JAR system" (Bullock et al. 1972b; their Fig. 11) offers at least two boxes where such differences might reside: the box named "activity command" (representing a heterogeneous command for EOD frequency changes of unclear origin) and the box named "limiter". The properties of the limiter would enable the adult

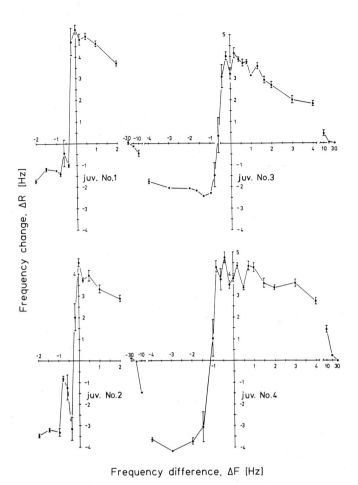

Fig. 4.52. Frequency difference vs response curves in *Eigenmannia lineata*. Like Fig. 4.49 but for four juveniles. Stimulus frequency unclamped, that is, constant throughout experiment. These juvenile fish gave good responses to ΔFs of both signs (although frequency increase was stronger than decrease). Strongest responses were evoked by ΔFs close to and including zero. Response curves are "off center" by ΔF ≈ −0.5 to −1 Hz (Kramer 1987)

females, for example, not to respond to +ΔF in spite of their responsiveness to −ΔF (Fig. 4.50).

A variable limiter should be provided for, because adult females occasionally gave JARs (frequency *increase*) to microprocessor-synthesized male EODs at +ΔF (Kramer 1985a). A variable limiter is also needed to provide for the presumed transitions of the juvenile to subadult, and subadult to adult response types (in part observed in adult female No. 3 = juvenile No. 5; Figs. 4.50 and 4.51).

While some kind of a limiter may explain the adult fishes' unresponsiveness to +ΔF (in adult males, any ΔF was almost or totally ineffective), no im-

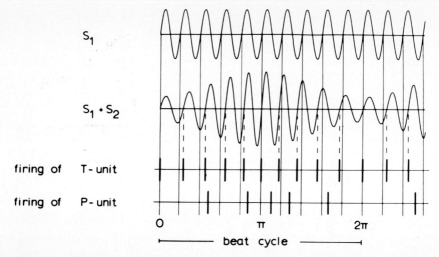

Fig. 4.53. Schematic representation of T- and P-receptor responses to a beat pattern resulting from the addition of two sine waves, S_1 and S_2, with S_1 being the dominant signal. For demonstration purposes, a beat pattern with a small number of carrier cycles was chosen. In the case of an EOD of 400 Hz and a ΔF of 4 Hz, 100 carrier cycles would fall into one beat cycle, and the full period, 2π, of the beat cycle would be 0.25 s long (Heiligenberg 1986)

mediate explanation is at hand for part of the juveniles' and one adult female's strong responses to $\Delta F = 0$ Hz (Figs. 4.52 and 4.50, respectively), a stimulus thought to be ineffective in previous studies. For $\Delta F = 0$ Hz, theory specifically predicts no JAR as there is no periodic variation in the combined signal (EOD superimposed by the stimulus). The presumed electrolocation performance, as studied by an overt, spontaneous following response to moving objects, was unimpaired under a jamming stimulus sufficiently close to $\Delta F = 0$ Hz (Matsubara and Heiligenberg 1978).

In theory, the fish's EOD serves as a kind of "carrier" frequency which is modulated by the stimulus signal in amplitude and phase at the beat frequency, that is, the frequency difference (Figs. 4.53, 4.58B). Electroreceptors of two types are sensitive to these periodic variations within a beat cycle: P-receptors (probability of firing) reflect the amplitude envelope of the beating field, and T-receptors (phase of the 1:1 spike) reflect the phase modulations of zero-crossings within a beat (Fig. 4.53). Hence both receptor types transmit information on ΔF. The fish could distinguish $+\Delta F$ from $-\Delta F$ by comparison of the time courses of amplitude and phase modulations of the beating field; true frequency analysis seems unlikely (Scheich and Bullock 1974; Scheich 1974, 1977a,b,c; review Scheich 1982; review Heiligenberg 1986).

The sign of ΔF could be determined from these electroreceptor responses in two ways: (1) by analysis of the amplitude envelopes of the beat patterns which, because of the harmonic content of the EOD, are time-asymmetric mirror images for identical ΔF values of opposite sign (Scheich 1974, 1977a,b,c), and (2) by comparing P- and T-receptor responses from different

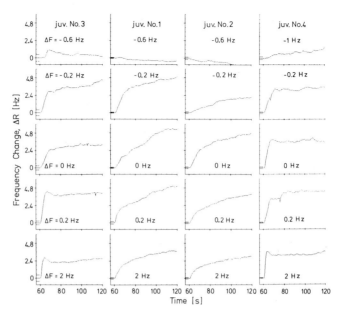

Fig. 4.54. Time course of frequency changes, ΔR, in response to stimuli of various ΔF values in the juvenile *Eigenmannia* of Fig. 4.52. Each curve represents the average from 12 repetitions. Stimulus frequency unclamped, that is, constant throughout experiment. Stimulus onset is at 60 s. Resting mean frequency (±1 SD, based on 120 measurements during the 60 s preceding stimulation onset) indicated at the left of each plot. Conditions: see legend to Fig. 4.49. At ΔF = −0.6 Hz (*top row*) fish were ambivalent in their choice of response direction, or did not respond at all (about −1 Hz in juvenile No. 4, *right column, top*). At ΔF ≥ −0.2 Hz (*second row from top*), only positive ΔRs were evoked (including ΔF=0 Hz). Note that the curves for ΔF=0 Hz, resemble those of positive ΔFs (*bottom two rows*), with no apparent "trial-or-error" behavior (Kramer 1987)

skin areas and detecting a kind of "motion" as specified in an amplitude-phase state-plane model, or Lissajous figure (Heiligenberg et al. 1978b; reviews Heiligenberg 1986, 1988). The motion in this model is of opposite direction for +ΔF and −ΔF. (Using an oscilloscope, the Lissajous figure method enabled Watanabe and Takeda (1963) to detect minute frequency changes of *Eigenmannia* to stimulus frequencies of small ΔF.)

An explanation of the juveniles' (Nos. 1–4, Fig. 4.52; and one additional juvenile male's) and one adult female's (No. 3, Fig. 4.50) responsiveness to ΔF=0 Hz might be that random fluctuations of their EOD rates (although very small, see Sect. 3.2.2.2) trigger the response. Once there is a frequency difference, however small, this would elicit the response. This explanation cannot be ruled out on the basis of the experiments of Figs. 4.49–4.52. However, the explanation seems insufficient, since the observed frequency changes were always in the same direction without detectable delay: frequency increase in the case of juveniles Nos. 1–4, but frequency decrease in female No. 3 and an additional juvenile male (Fig. 4.54). Frequency change in both directions and delay, at least occasionally, should be observed if the "random frequency fluc-

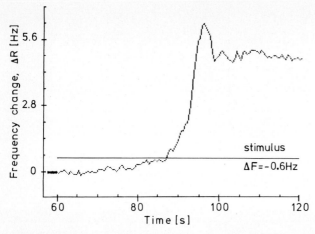

Fig. 4.55. Frequency change, ΔR, of a juvenile *Eigenmannia* (No. 3 of Fig. 4.52) over time in response to ΔF = −0.6 Hz (single experiment, not averaged; otherwise like Fig. 4.54). Stimulus frequency (*horizontal line*) unclamped, that is, constant throughout experiment. Stimulus onset is at 60 s. Although the fish should lower its frequency in order to increase the difference, it slowly raised its frequency and, after more than 25 s, crossed over it. Only then was a JAR evoked (although of opposite sign). Conditions: see legend to Fig. 4.49 (Kramer 1987)

tuation hypothesis" was to be applied. This has been observed in the unrelated African mormyriform fish with a similar wave discharge, *Gymnarchus niloticus*, which, in contrast to *Eigenmannia*, commonly shows irregular fluctuations in EOD frequency and "singing" (Bullock et al. 1975; their Figs. 2A–C, 3), and spontaneous and stimulation-evoked discharge stops (Lissmann 1958; Szabo and Suckling 1964).

Juveniles Nos. 1–4 showed the behavior expected at ΔF=0 Hz only at ΔF = −0.6 Hz to −1 Hz (Fig. 4.52): the fish (a) delayed choosing the sign of their response or (b), often in addition, chose to change their frequency in the wrong direction (Fig. 4.55) or (c) did not respond at all (|ΔF|<0.3 Hz).

Does this mean the fish were unable to determine the correct sign of ΔF, although at ΔF = −0.6 Hz there was one beat cycle per 1.67 s? (One-quarter of a beat cycle, 0.42 s in this case, is sufficient for the fish to determine ΔF; Bullock et al. 1972b.) How could the fish quickly and consistently respond to stimuli of ΔF=0 Hz in the absence of a detectable frequency difference? Occasionally, the stimulated fish still changed their frequency in the "wrong" direction, even at ΔF = −1.5 Hz (one beat cycle per 0.67 s). From the uncertainty about which sign the response would take at ΔF = −0.6 Hz, the accuracy of ΔF assessment is not better than ±0.3 Hz.

At ΔF=0 Hz, cues not yet identified must enable the fish to respond in a predictable way. This is not to deny that once a certain threshold frequency difference is reached, the response might be maintained by amplitude and phase modulations of the combined field beating at increasingly higher frequencies as the fish continues to change its frequency.

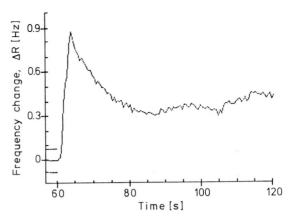

Fig. 4.56. As Fig. 4.55, but for juvenile No. 1 (of Fig. 4.52). A biphasic square wave stimulus was electronically phase-locked to the fish's EOD, that is, the stimulus was clamped at a constant $\Delta F = 0$ Hz throughout the experiment. Stimulus onset is at 60 s. Average curve from 12 experiments. Note strong frequency change although there was no periodic modulation of the EOD by the stimulus. Phase difference: 75°; stimulus intensity, 30 µV · cm^{-1} (Kramer 1987)

A JAR can even be evoked in the maintained absence of beating of the fish's EOD field, by the use of a frequency clamp set at $\Delta F = 0$ Hz (Fig. 4.56). A square-wave, triggered cycle per cycle by the fish's EOD, evoked a strong and immediate frequency increase, but only at certain phases relative to the EOD. In this experiment, ΔF was strictly maintained at 0 Hz during the fish's response.

One possible cue enabling the fish to respond to $\Delta F = 0$ Hz is a noise-like pulse of broad spectral composition associated with sudden stimulus onset, although this could give information only about the time of stimulus onset, and not about ΔF. It might still cause the fish to give a frequency change (even a systematic one) and thus trigger the JAR. This explanation is unlikely, however, because of the "soft" stimulus onset (rise time 400 ms) and the low stimulus intensity used in these studies (Kramer 1985a, 1987).

A more likely cue is a step-like change of the fish's perceived EOD amplitude at stimulus onset (at least in those parts of its skin where the stimulus gradient is maximal). This hypothesis is supported by the observation of frequency drops of up to 4.5 Hz in response to sudden changes of environmental resistance (hence perceived EOD amplitude) in *Eigenmannia* (Larimer and MacDonald 1968, their Fig. 9). A step-like amplitude change of the EOD is also caused by superposition with a wave-like signal of identical frequency.

An amplitude change should be perceived by P-receptors; T-receptors should perceive the change of phase of the voltage gradient crossing the zero line and might respond even if just the slope of that gradient changed without any change of phase.

The amplitude change caused by a stimulus of $\Delta F = 0$ Hz can be an increase or a decrease, depending on the phase difference, and should lead to opposite changes of the response patterns of P- and T-receptors. Stimulus phase was not controlled in the experiments of Figs. 4.49–4.52 (but in those of Fig. 4.56); therefore, stimuli of all phases were probably used. In spite of this, there was only one response pattern per fish (Fig. 4.52; female No. 3 in Fig. 4.50; and an additional juvenile male). Phase-locked stimuli of all phases had been found ineffective by Watanabe and Takeda (1963), Bullock et al. (1972b) and Heiligenberg et al. (1978b). This is probably explained by the very small number of fish studied without reference to sex or age groups.

An alternative, or additional explanation of the sensitivity of a group of juveniles and of female No. 3 (Fig. 4.50) to stimuli of $\Delta F = 0$ Hz, involves the ampullary electroreceptors with their acute low-frequency sensitivity (reviews by Bennett 1971b; Szabo and Fessard 1974; Bullock 1982; Zakon 1986). A step-like EOD amplitude change should be clearly sensed by these receptors. The suggestion of Bullock et al. (1972b) that ampullary receptors might play a role in the JAR by sensitivity to the low-frequency beat envelope has not been investigated. Bastian (1987a,b) has found evidence suggesting that ampullary receptors can provide supplementary information for active electrolocation in the presence of jamming signals in *Apteronotus leptorhynchus*, although he does not specify whether a JAR occurred or not.

Functions of the Divergent Frequency Change Behavior. Although the JAR has never been observed in nature, there exist a few laboratory observations of two-fish interactions (Bullock et al. 1972a,b; Fig. 4.48), and experiments on the effect of artificial, jamming stimuli on an unconditioned, overt following behavior presumably mediated by the electric sense (for reviews see Heiligenberg 1977, 1986). Heiligenberg concludes that the JAR is part of an "early warning system", enabling the fish to shift to a "safer" frequency, long before an approaching intruder with similar EOD frequency can disrupt its electrolocation (but see below).

EOD frequency modulations during social behavior have been described in *Eigenmannia virescens* (Hopkins 1974a; Hagedorn and Heiligenberg 1985; see previous chapter). The time course of one kind of modulation, the "long rise", resembles a JAR given to $+\Delta F$. Is the reason adult *Eigenmannia* do not give JARs to $+\Delta F$ the similarity of that response to the long rise signalling submission or retreat? (Adult females stimulated with computer-synthesized male EODs of $+\Delta F$ sometimes gave a $+\Delta R$; Kramer 1985a).

The reason adult males do not (or only weakly; Fig. 4.49) respond to $-\Delta F$ is not known; in nature, adult males are unlikely to give JARs placidly on encountering a conspecific with jamming EOD frequency, as they are extremely aggressive, chasing away conspecifics. Vicious fights are the rule, especially with other large males, followed by prolonged, high-intensity chasing of the loser by the winner, so that the fish have to be separated (360-l aquaria of 120×60 cm bottom area, richly planted and with many shelters). The males of the present study often gave short rises and interruptions in response to the jamming signals, categories of transient EOD frequency modulations observed in threatening fish likely to attack (Hopkins 1974a).

The chance that adult males might meet conspecifics of similar frequency (except large males) seems low, since the frequencies of all four adult males

(>30 cm) were at the low end (268–364 Hz; Kramer 1985a) of the species' frequency distribution (260–650 Hz at 27°C for *E. virescens* which is easily confused with *Eigenmannia lineata*; Hopkins 1974a; Westby and Kirschbaum 1981). The EOD intensity of adult males (100 mV$_{p-p}$ head-to-tail, or more) is so much higher than that of the other sex or age groups (up to 16 mV$_{p-p}$; Kramer 1985a) that these fish, on meeting an adult male of suitable EOD frequency, would give a response long before the adult male (for stimulus intensity vs response relationship see Kramer 1985a).

It is unlikely, however, that adult males would elicit JARs in adult females, as these only respond to frequencies higher than their own (Fig. 4.50). If, however, the male happenend to display a frequency closely above half the female's a JAR would be elicited, as the strong second harmonic of the male EOD would be close to the fundamental frequency of the female's EOD (Kramer 1985a). A male discharging at half the female's frequency is probably not rare in nature (for evidence see Kramer 1985a). Whether such a response would be of social significance is impossible to ascertain from our present knowledge. *Eigenmannia* does, however, discriminate synthetic male from female resting EODs in the absence of frequency or amplitude cues (Kramer and Zupanc 1986).

The behavior of all four groups of individuals, especially adult fish, appears more or less maladapted, because a symmetrical frequency difference vs response curve with strongest responses for smallest ΔFs (which no group showed) would be optimal for electrolocation in the presence of jamming stimuli. It is doubtful, however, whether electrolocation performance often suffers under natural conditions when a JAR would be elicited. Presumed electrolocation performance, as measured by a spontaneous, unrewarded following behavior to moving objects, deteriorated only when the intensity of a jamming stimulus approached the fish's own near-field EOD intensity (Heiligenberg 1977). To experience such high intensities from another fish's EOD field, two fish of comparable size (and almost identical EOD frequency) must be very close to each other (Heiligenberg 1977 gives 4 cm; his Fig. 34). From aquarium and field observations, such instances are rare (except during fighting or courtship, when a JAR is unlikely because of the presence of other kinds of frequency modulations; see above).

The electrolocation performance of *Sternopygus* males, which did not show a JAR, was only impaired at an unphysiologically high stimulus intensity of 50 times their own near-field intensity (2.5 mV · cm^{-1}; Matsubara and Heiligenberg 1978). This immunity of *Sternopygus* to jamming stimuli gave rise to an alternative hypothesis of how an electric fish might detect moving objects in the presence of jamming stimuli (briefly reviewed in Heiligenberg 1986): the spatial pattern of amplitude modulations of the fish's own EOD caused by small, moving objects certainly differs from the more global ones caused by a distant dipole current source, such as a conspecific. This difference might enable the fish to distinguish between the two kinds of disturbances. This hypothesis might well also apply to *Eigenmannia*, the adult males of which did not show a JAR (or only one considered to be too small to be called a JAR in *Sternopygus*; see above).

Also, the observation of strong JARs (Figs. 4.50, 4.52) to stimuli of $\Delta F = 0$ Hz (which do not impair electrolocation performance; Matsubara and Heiligenberg 1978) in some fish, of strong habituation, and of a disconcertingly high inter- (and sometimes intra-) individual (and probably geographical, see Kramer 1987) variability in all fish do not suggest that the JAR is important in active electrolocation (object detection).

This shows that we know little about the function(s) and not enough about mechanisms of the JAR (see previous section of this chapter). We can only speculate about the selective forces which shaped the response, such as those probably imposed by the other function of the electric system, communication. However, I suggest that the name of JAR should not be changed, until the true function has been securely established. Watanabe and Takeda's (1963) original assumption that the JAR may serve the fish "to distinguish between its own signal and those of its neighbors" for better object detection may be very close to truth; as a hypothesis, I would like to replace the qualifier "for better object detection" by "for better social identification including location of individual conspecifics, and more effective, individualized com-

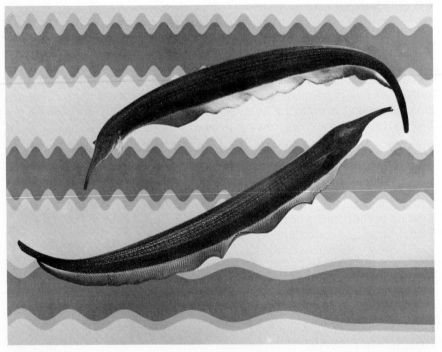

Fig. 4.57. Social encounter of two *Sternarchorhamphus*. The screen in the background shows the beating of the electric fields generated by the two fishes' electric organ discharges. The sequence of the traces is from top to bottom. The beat frequency slows down and finally the amplitude of the trace reaches a stable value. This indicates phase coupling. Higher harmonics of the electric organ discharges were filtered out so that the beat envelope is sinusoidal. At the start of the top trace the beat frequency is 5 Hz. The fish measure about 30 cm (Langner and Scheich 1978)

munication". In those fish usually displaying a constant frequency, the function of the JAR seems to be the facilitation of effective communication between members of a school, rather than the facilitation of object detection (see also Kramer 1990).

The JAR could have acquired functions other than the ones it originally served (through evolutionary change as described by the ethological concepts of ritualization and emancipation; see, for example, Manning 1979), as it could clearly have been a preadaptation (see, for example, Wilson 1975) for social signalling by frequency modulations. Investigations into the social behavior and individual life histories should help clarify some of the obscure points.

Active Phase Coupling. Pairs of the apteronotid fishes *Sternarchorhynchus* sp. and *Sternarchorhamphus* sp. of very high EOD frequency were sometimes found to actively *reduce* their frequency difference, synchronizing their wave EODs at identical frequencies for a few s up to 4 min (longer bouts with stronger stimuli) (Fig. 4.57; Langner and Scheich 1978). The initial ΔF may be 1–20 Hz; a slow "approach" phase by a frequency increase was followed by a sudden locking of one fish onto the EOD of the other (or an artificial stimulus) at a preferred phase relationship. Phase jitter may be as low as a few

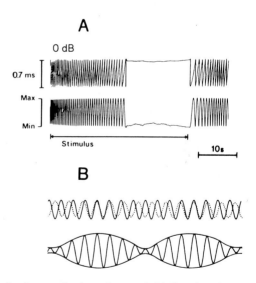

Fig. 4.58. A. Active phase coupling in *Sternarchorhamphus* sp. Initially, the stimulus frequency was a few Hz lower than the EOD frequency. The *lower trace* is the time course of the beat envelope; the *upper curve* shows the concurrent phase oscillations of the fish's discharge within the stimulus cycle (about 0.7 ms = 360°, or a full cycle, at a stimulus frequency of 1408 Hz). During stimulation the beat frequency slowed down, that is, the fish actively reduced the frequency difference. In this instance, phase coupling, or frequency identity, occurred while the beat amplitude was close to its minimum. *B* Illustration of a beat pattern with amplitude envelope (*below*) and of the two constituent frequencies (*above*). The two wave trains go in and out of phase with a period which corresponds to the reciprocal of the frequency difference. Upper solid line of amplitude envelope corresponds to beat amplitude as shown in lower trace of *A* (Langner and Scheich 1978)

µs over half a minute, especially with a strong stimulus (such as the EOD of a nearby conspecific; Fig. 4.58). Phase coupling was abandoned as suddenly as it began by a further frequency increase, with the EOD frequency crossing over the stimulus frequency (up to a final frequency difference of 15 Hz).

The behavior of active phase coupling is the opposite of the jamming avoidance response: two fish differing in EOD frequency by a few Hz do not try to have the beat frequency resulting from the addition of their electric fields increased but slow it down until synchronization occurs.

Phase coupling also occurred to the EODs of other species and to artificial stimuli even several hundred Hz lower in frequency than the EOD frequency, although no frequency identity was achieved. Due to the fish's frequency increase by a few Hz, standing wave patterns on the oscilloscope (meriting the name phase coupling) occurred when the stimulus and the EOD frequency were higher harmonics of a common fundamental frequency, and phase coupling was to every nth stimulus cycle only. Stable phase relationships not in every cycle but in every nth discharge cycle are found when both frequencies stay in a harmonic relation $n{:}m$ (1:2, 1:3, ..., 2:3, 2:4, 2:5, ...). In one observation, every 11th EOD of a *Sternarchorhynchus* discharging at 1346 Hz after increasing its resting frequency by 12 Hz, was phase-locked to every tenth stimulus cycle of $f = 1224$ Hz.

The biological significance of phase coupling is not yet clear. Langner and Scheich (1978) discuss several possibilities: (1) Phase coupling may be an alternative strategy of jamming avoidance, as a $\Delta F = 0$ Hz situation does not interfere with object detection, similar to a high beat rate situation, (2) phase coupling might be a new social signal. This is suggested by the observation of male and female *Sternopygus macrurus* engaging in a 1:2 frequency relationship (Hopkins 1974b) which may involve active phase coupling (not yet shown), and (3) phase coupling could be a way of electrical hiding, because mutual localization may become more difficult when the two electrical dipole fields do not beat.

Although less dramatic than in *Sternarchorhynchus* and *Sternarchorhamphus*, phase coupling has been found in several other species as well: *Eigenmannia (humboldti?)*, *Sternopygus sp.*, *Apteronotus (hasemani?)*, *A. leptorhynchus*, and in the pulse fish *Rhamphichthys (rostratus?)* (Gottschalk and Scheich 1979). The response of the latter fish may resemble the synchronization behavior in *Gymnotus carapo* (Sect. 4.1.2.1), and even the unrelated mormyrids exhibit a similar behavior, the preferred latency (or echo) response (see Sect. 4.1.1.1).

Stimulus Filtering and Sex Recognition in Eigenmannia. Like other lateral line receptors (review Hudspeth and Corey 1977; Hudspeth 1985), tuberous electroreceptors of weakly electric teleosts exhibit tuning to a specific stimulus frequency, and filter properties (see Chap. 2; Scheich et al. 1973). The frequency that evokes the strongest response from a receptor unit, or "best" frequency, usually is close to the EOD frequency for that fish (Fig. 4.59; Hopkins 1976).

However, behavioral frequency-threshold curves are either more complex than the generally V-shaped receptor tuning curves, or deviate quantitatively

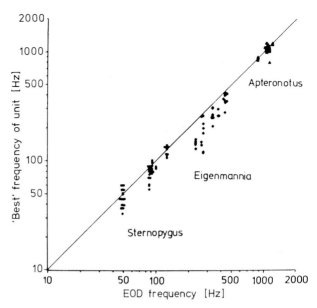

Fig. 4.59. Graph shows the tuning of tuberous electroreceptors of three species of wave gymnotiforms to an individual's EOD frequency. The *diagonal line* shows perfect correspondence between EOD frequency and best frequency of the tuning curve (After Hopkins 1976 from Scheich 1982)

from them (Fig. 4.60; Knudsen 1974). This shows that considerable central integration is involved in these fishes' sensory mechanisms. In the context of the perception and analysis of an imposed sine wave stimulus, there could be, and probably is, central sensory integration of (1) different types of electroreceptor afferences; (2) integration over the variation in best frequencies within a tuberous receptor type (see below); (3) extraction of temporal features of the beating field, resulting from the combination of EOD and stimulus (see Sect. 4.2.2.2).

The best frequency of a tuberous receptor type may exhibit considerable variation within the same individual (Viancour 1979a), and may vary according to local changes of the EOD amplitude spectrum along a fish's body (Bastian 1977). Filter properties of tuberous receptors differ among species and types of tuberous receptors; the filter's high frequency cutoff slope may be 70 dB/octave (more typically: 40 dB), and 30 dB/octave (more typically: 10 dB) for its low frequency cutoff slope.

Receptor filter properties may suppress the EODs of other species with markedly different frequencies (for example, those shown in Fig. 4.59). A fish's receptor tuning to its own EOD should improve its signal-to-noise ratio in both communication and object detection. It is unlikely, however, that filter mechanisms of receptor tuning aid significantly in discriminating between different individuals of the same species. Filter cutoff slopes are often not very steep, nor are EOD frequencies of one's neighbors very different from one's own.

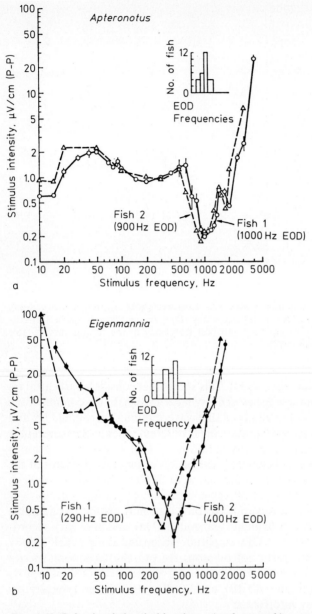

Fig. 4.60. Behavioral threshold voltage (peak-to-peak) versus stimulus frequency of trained fish in a food-rewarded, modified T-maze paradigm. *a* Two *Apteronotus*, one with an EOD frequency of 1000, the other of 900 Hz. *Standard error bars* only shown for one fish. *Insert* histogram of laboratory-held *Apteronotus* with a given EOD frequency at 26°C. *b* as *a*, but for two *Eigenmannia* of 290 and 400 Hz EOD frequency. Note keen low-frequency sensitivity in *Apteronotus* which is absent in *Eigenmannia*. Both fish are most sensitive to frequencies within about 50-100 Hz of their EOD frequencies (Knudsen 1974)

In no electric fish species is it known whether *spectral* or *temporal* properties of tone-signals, including *Eigenmannia's* sexually dimorphic EOD, provide the cues evoking a behavioral response (frequency domain vs time domain mechanisms; see also Scheich 1977b, p. 223).

More complex sensory mechanisms (like those subserving the jamming avoidance response and active phase coupling; see Sect. 4.2.2.2) could be involved for that job. Substantial support for this view was the discovery of a wave fish community exhibiting a most extensive degree of overlapping of EOD frequencies (Kramer et al. 1981a; see Sect. 3.2.2.2).

A possible way for wave fish to recognize conspecifics and mates would be sensitivity to the finer detail of their wave EOD which is species-specific, or, as in the genus *Eigenmannia*, specific for a very small number only of closely related, sympatric species that are exceedingly difficult to distinguish (see Sect. 3.1.2.2). (The latter fish need, of course, additional mechanisms to prevent mixing of their gametes; to date, these mechanisms are unknown, as the taxonomy of the genus *Eigenmannia* is confused). Digital synthesis of natural EODs (Kramer and Weymann 1987) allowed the study of behavioral sensitivity for the differences between male and female EOD waveforms in *Eigenmannia*.

Three experimental approaches were chosen: (1) the jamming avoidance response was used as a tool for exploring the fishes' sensitivity for the natural variability of the species' EOD waveform, and of artificial stimuli; (2) with the same intention, trained fish were used in a conditioned discrimination paradigm; (3) a spontaneous preference test was used in order to gain information about biological significance of the EOD waveform variability, and as a control for experiment (2). Surprising differences in the results were found between (1) on one hand, and (2) and (3) on the other, shedding light on *Eigenmannia's* complex sensory capacities which seem to depend on behavioral context.

Principles of Wave Analysis as Revealed by JAR Experiments. This part of the study aimed at finding out whether differences in waveform or harmonic content of stimuli are perceived by the electrosensory system, as estimated by the JAR. The situation as found in the literature was far from clear.

According to Bullock et al. (1972b), comparison of the effectiveness of square, sinusoidal, triangular and sawtooth waveforms in eliciting JARs showed "very little difference" in stimuli close to the EOD frequency ($\Delta 1F$ stimuli); unfortunately, no data were presented. At one-half EOD frequency ($\Delta \frac{1}{2} F$ stimuli), however, only the sawtooth waveform with its strong second harmonic (which was close to the EOD frequency) elicited JARs.

It is difficult to understand why sensitivity for differences in waveform or harmonic content should be detectable in $\Delta \frac{1}{2} F$ stimuli only, but not in the behaviorally more important $\Delta 1F$ stimuli. It was not possible to draw conclusions from this and other conflicting evidence (reviewed in Kramer 1985a), because information about the stimuli (waveforms and Fourier amplitude spectra) and the strengths of responses was lacking or incomplete.

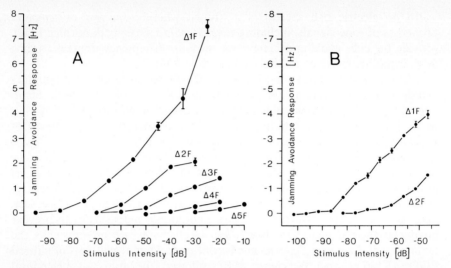

Fig. 4.61. Response as function of intensity of a sine wave re: 29.6 mV$_{p-p}\cdot$cm^{-1} (0 dB). Each point is a mean of at least 10 measurements. *Standard errors* are either shown or are too small to be drawn. *ΔIF curves* stimulus frequency was 2 Hz below (*B* above) the EOD baseline frequency (first harmonic). *A Eigenmannia* sp. 3. *Δ2F* curve stimulus frequency was 2 Hz below two times that frequency (the second harmonic); and so on. *B E. lineata*. *Δ2F curve* stimulus frequency was 4 Hz above the second harmonic of the EOD (Kramer 1985a)

Therefore, stimulation experiments were performed in which both stimulus and responses were controlled or measured quantitatively. The response strength as a function of intensity of a sine wave had to be determined first, because there was some disagreement in the literature as to how both were related (reviewed in Kramer 1985a).

The JAR is approximately proportional to the logarithm of stimulus intensity over a range of at least 40 dB (Fig. 4.61), as already suggested (on the basis of few data) by Watanabe and Takeda (1963). Therefore, an exponential function, $y = a^x$, best approximates most of the relationship between stimulus intensity, x, and response strength, y. This input-output relationship is found in most sense organs, and is known in psychophysics as the Weber-Fechner Law. Within certain limits, a given percentage of change in stimulus intensity evokes the same increment in response strength, independent of absolute stimulus intensity.

The JAR is also elicited by sinusoids close to low integer mutiples, or higher harmonics, of the EOD frequency (Watanabe and Takeda 1963), with response strengths decreasing the higher the harmonic (up to the fifth) at constant stimulus amplitudes (Bullock et al. 1972b). These authors also found that the best (frequency-clamped) ΔF remains at about the same value from the first to the fifth harmonic.

Stimulus/response curves of higher harmonics are of sigmoid shape; no depression of the responses even at very high intensities is observed when

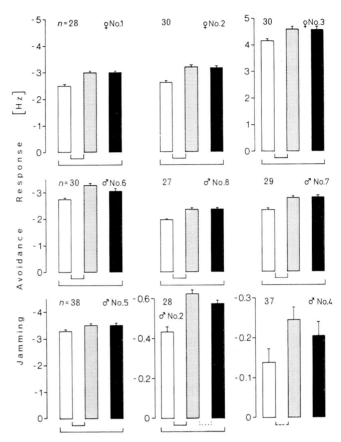

Fig. 4.62. Jamming avoidance responses, shown as means and standard errors, to computer-synthesized stimuli of three waveforms of equal amplitudes (p−p). These were the male EOD *(white)*, the female EOD *(stippled)*, and sine waves *(black)*. Significant differences of responses, as revealed by paired statistical tests, are indicated by *square brackets, $P<0.001$; broken line bracket, $P<0.01$; dotted bracket, $P<0.02$. N* Number of experiments per waveform. Note that the male EOD evoked weaker JARs than both the sine wave and the female EOD with stronger f_1 harmonics (Kramer 1985a)

inter-trial intervals are sufficiently long so that habituation is avoided (Fig. 4.61).

The effectiveness of various stimulus waveforms of equal peak-to-peak amplitudes, and of frequencies close to the fish's EOD frequency (Δ1F-stimuli), were compared. All tests showed that the response strength was proportional to the intensity of the fundamental frequency or first harmonic, f_1, of a stimulus waveform.

This was shown by comparing the responses to: sine waves (the most effective stimulus), square and sawtooth waves (undistorted and those distorted by an isolating transformer which introduced frequency-dependent phase shifts), synthesized complex waves consisting of a sine wave with its f_2-harmonic added at specified phase differences, and synthesized male and female EODs

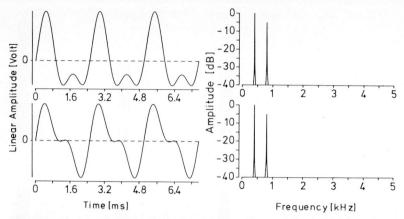

Fig. 4.63. Computer-synthesized stimulus waveforms as recorded from the water, *left*, composed of two superimposed, harmonically related sine waves. To a sine wave of fundamental frequency, f_1, a sine wave of twice that frequency, f_2, of weaker amplitude (−5 dB) was added. *Right* Fourier amplitude spectra. *Ordinates* of left diagrams are arbitrary linear amplitudes (*V*), of right diagrams amplitudes expressed as dB attenuation relative to the strongest spectral component of each waveform. *Top* Phase difference between the peak amplitudes of the two harmonics is 0° (no phase difference). *Bottom* Phase difference is 90° (or a quarter cycle). Note that the spectral amplitudes of both waveforms are identical (Kramer 1985a)

indistinguishable from natural ones (all could be manipulated on-line to an experiment in amplitude, frequency, and rise/fall times).

Average JARs, obtained with a Δ1F stimulus of any waveform of given f_1 intensity, were not statistically different from expected JARs as calculated for a sine wave of that f_1 intensity, presented at the same ΔF.

For example, at equal peak-to-peak amplitudes, the f_1 intensity of the synthesized EOD of a large male was only 71% of that of an adult female's, which were both used as stimuli in these experiments. Male EODs elicited correspondingly weaker responses than female EODs (a mean difference of 19%, Fig. 4.62; $P<0.01$ in each of nine fish tested). The responses to female EODs and sine waves of identical f_1 intensities were so similar that the differences were not significant. (Only in large male No. 2 was a difference observed: his responses to female EODs were even stronger than those to sine waves. Similar to other large males, his responses were very weak; see Fig. 4.49.) The mean difference between the responses to male and female EODs agrees with the expected difference as calculated on the assumption that the f_1-intensity determines the reponse alone (differences Observed-Expected JARs non-significant; Kramer 1985a). The human ear not only discriminates audio playback of synthesized male and female EODs from each other, but also both EODs from a sine wave audio signal.

These results suggested a mode of wave analysis which focuses exclusively on the f_1 intensity of a Δ1F stimulus. This was confirmed by the use of three synthesized waves of equal f_1 intensities. Two were sine waves which had an f_2-harmonic added at phase angles of 0° and of 90°, so that their amplitude spectra were identical although their waveforms were not (Fig. 4.63); the third was was a pure sine wave. The intensity of the f_2 harmonic was that of the f_2 harmonic of a large male's EOD (-5 dB of the fundamental frequency, f_1).

These waveforms were calculated according to:

$$y = \sin \omega t + a \sin (2\omega t + \varphi),$$

where $\omega = 2\pi f$, $f=$ frequency, $t=$ time, $a=$ amplitude, $\varphi=$ phase difference relative to peak amplitudes.

Unlike the symmetrical waveform with a 90° phase difference, a phase difference of 0° leads to an asymmetrical waveform with respect to intervals between zero-crossings (similar to the male EOD). The intensities of the two signals' f_1 components were identical to that of the sine wave which was used for comparison and scaling; hence, the three signals differed in peak-to-peak amplitudes. These amplitudes were 1.00, 1.18 and 1.37 for the sine wave, the 0°- and the 90°-waveform, respectively. The relative energy contents, as determined by $V_{r.m.s.}$-measurements, were 1.00 for the sine wave and 1.15 for the other two waves.

Six *Eigenmannia lineata* (three of each sex) were insensitive to these differences in phase relationships, waveform, peak-to-peak amplitudes, and power, as determined by the JAR (all $P > 0.05$; Kramer 1985a, his Fig. 10). As in all other experiments, response strength appeared to be exclusively determined by the intensity of the f_1 harmonic (or fundamental), which was the only parameter in common to all three signals (besides, of course, ΔF). The human ear can immediately distinguish an audio sine wave signal from the other two waves, but cannot discriminate between the latter two.

The mode of signal analyis as suggested above was further tested by using stimuli of close to one-half or one-third EOD frequency. Bullock et al. (1972b) and Heiligenberg et al. (1978b) had found that certain subharmonic stimuli, excluding sine waves (Watanabe and Takeda 1963), do evoke JARs, but quantitative data, and a coherent explanation of their effectiveness, were lacking. Such data are, however, needed for testing models of signal analysis, like the one outlined above.

As expected, and graded with intensity, $\Delta\frac{1}{2}F$ sawtooth stimuli evoked strong JARs, while the square wave (and, of course, the sine wave) were ineffective (Fig. 4.64; see also Kramer 1985a, his Table 2). Only the sawtooth has a strong second harmonic, f_2, close to the EOD frequency (Kramer 1985a, his Fig. 8). At $\Delta\frac{1}{3}F$, however, both the sawtooth and the square waves proved effective stimuli; the sine wave again was not. Both effective stimuli have a strong third harmonic, f_3.

Therefore, in these experiments with subharmonics it appeared that, for a given ΔF, the JAR was determined by the intensity of that higher stimulus harmonic which was close to the fundamental frequency f_1 of the EOD, irrespective of stimulus waveform and subharmonic frequency.

Two further experiments confirmed this. The two computer-calculated waves of equal amplitude spectra but different waveforms (Fig. 4.63) evoked similar JARs not significantly different from each other when presented at equal intensities of their harmonics (as discussed above), at frequencies close

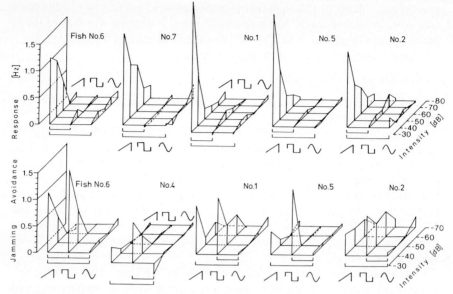

Fig. 4.64. Jamming avoidance reponses of *Eigenmannia* sp. 3 to subharmonic stimuli of three waveforms of equal amplitudes (peak-to-peak): sine, square and sawtooth. Each point is the mean of 10 experiments. The *z-axis* is the stimulus intensity re: 29.6 mV$_{p-p}$ · cm^{-1}. *Top* The stimulus frequency was 2 Hz below one-half EOD frequency. *Square brackets* indicate significant response differences, as shown by paired statistical tests, for the highest stimulus intensities ($P<0.001$). Note that JARs were elicited only by the sawtooth wave with its strong second harmonic. *Below* Stimulus frequency was 2 Hz below one-third EOD frequency. *Square brackets* as above, but $P<0.05$. JARs were evoked by the square and the sawtooth waveforms which both contain strong third harmonics (Kramer 1985a)

to one-half that of the fish's EOD (Kramer 1985a; his Table 3). Although synthetic male EODs had been less effective than synthetic female EODs of equal peak-to-peak amplitudes at Δ1F stimulus frequencies (Fig. 4.62), their relative effectiveness reversed when used as Δ½F-stimuli: the male EOD with its strong second harmonic evoked a much stronger response than the female EOD (Kramer 1985a, his Table 4). (The second harmonic of the male EOD was stronger than that of the female EOD by a factor of almost 4 in terms of the f_2/f_1 ratio).

All experiments showed that it was not the shape of the waveforms which determined the strength of the JAR. Signals were effective according to the intensities of their f_1 components (in a Δ1F stimulus), not peak-to-peak amplitudes, power nor waveform. Experiments with subharmonic stimulus frequencies showed that the effective harmonic of the stimulus need not be its fundamental frequency, as suggested earlier by Bullock et al. (1972b) for a Δ½F sawtooth wave. Responses to Δ½F and Δ¹/₃F stimuli of nine waveforms, including synthesized EODs, were obtained only when the signal contained a harmonic close to the EOD frequency which was sufficiently strong. That harmonic was, of course, the f_2 component in a Δ½F stimulus, and the

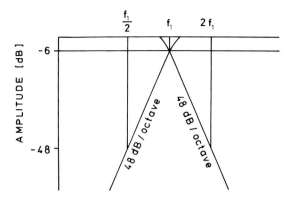

Fig. 4.65. Spectral line filtering by an electronic band-pass filter with high and low cutoff frequencies adjusted to the same frequency, f_1. Note that any higher or lower harmonics of a stimulus wave are suppressed, except the harmonic that is sufficiently close in frequency to f_1

f_3 component in a $\Delta^1/_3 F$ stimulus. Response strength was correlated with the intensity of a single harmonic, and not significantly different from that expected on the assumption of sharp spectral line filtering, as achieved by an electronic bandpass filter with high and low cutoff frequencies adjusted to the same frequency (Fig. 4.65). Further support was the observation of identical responses to two waves (consisting of an f_1 with added f_2-component) very different in form (or phase relationships between harmonics), but identical in terms of Fourier amplitude spectra and intensities of their constituent harmonics, when presented as $\Delta 1F$ or $\Delta \frac{1}{2}F$ stimuli.

An especially illuminating result was obtained with the synthesized "natural", sexually dimorphic EOD waveforms. Their effectiveness reversed with stimulus frequency. Female EODs were more effective than male EODs when presented as $\Delta 1F$ stimuli, but much less effective than male EODs when presented as $\Delta \frac{1}{2}F$ stimuli, according to their intensity differences in f_1 and f_2 components.

What can be inferred regarding sensory processing from these behavioral experiments? First, the stimulus wave clearly appears to be bandpass-filtered and analyzed harmonically before mechanisms of time domain analysis can come into play. The fish were able to extract the intensity of any harmonic (at least up to the third) close to their EOD frequency from any waveform, whether the frequency was subharmonic or not. This is what should be expected from the known properties of electroreceptors (T- and especially P-receptors) which are approximately tuned to the EOD frequency, resembling broad bandpass filters (Scheich et al. 1973; Hopkins 1976; Viancour 1979a,b). These properties also explain why the strength of the JAR is progressively reduced for frequencies that are two, three or more times the EOD frequency. Second, the present results show that fish are insensitive to the waveform of a stimulus although waveform-dependent information is present in a beat (Scheich 1977a,b). This second conclusion, of course, follows from the first. Different waveforms of identical spectral amplitudes elicited similar responses, although the resulting beat patterns must have been quite different.

Fish apparently do not (or cannot) use all of the information theoretically available. The waveform-dependent information would become accessible only after exceedingly complex computation, as the local waveform of a fish's own EOD varies considerably along the fish's body (wave-discharging apteronotids: Bennett 1971a; Hoshimiya et al. 1980; Bastian 1981; pulse-discharging gymnotids: Bastian 1977). The interaction of an *Eigenmannia* male's and female's EOD with, for example, a sine wave of identical amplitude and frequency difference also produces quite different beats, reflecting differences in waveforms of the fishes' own EODs only.

In summary, the JAR experiments do not support the idea that fish can discriminate the differences between male and female EODs, for sharp bandpass-filtering seems to eliminate all sex-specific features from the stimulus. However, the fish do possess this sensory capacity, as later demonstrated in a different experimental context:

Conditioned Discrimination of Stimulus Waveforms. Analysis of a wave signal by sharp bandpass (spectral line)-filtering, as concluded in the above paragraphs, would give *Eigenmannia* all the information it needs, *and no more*, to successfully perform a jamming avoidance response to any stimulus waveform with a strong harmonic close to the fish's EOD frequency (how the fish estimates ΔF is a separate problem; see Sect. 4.2.2.2). However, for discriminating between conspecifics whose EOD waveforms differ according to sex and age (see Sect. 3.1.2.2, Fig. 3.15) bandpass filtering clearly is disadvantageous, since all individual-specific waveform information is lost. Should the fish possess a second sensory mechanism, operating in parallel, providing the fish with information about the stimulus waveform?

This has, in fact, been demonstrated in trained fish by the method of discrimination conditioning, and in spontaneous choice tests (using untrained fish; see below). *E. lineata*, trained to receive a food reward near a dipole presenting one signal waveform, hesitated to approach the dipole when it presented another (which was associated with mild punishment during the previous training period, should fish try to get a reward). In the critical tests, response latencies to the previously rewarded signals were significantly shorter than those to the previously unrewarded signals, everything being equal except stimulus waveform. *Eigenmannia* not only discriminated synthetic female from male EODs, but also sine waves from sawtooth waves, and sine waves from male EODs (all signals of equal peak-to-peak amplitudes; Fig. 4.66). The fishes' behavior changed appropriately when the roles of rewarded and unrewarded signals were reversed.

The most parsimonious explanation for this discrimination capacity might still be that found for the JAR: sharp band-pass filtering (Fig. 4.65). By that type of signal analysis fish might discriminate between any pair of signals, as shown in Fig. 4.66, by their differences in fundamental frequency intensities (compared to the sine wave of relative amplitude 0 dB, the fundamental frequency components of the other waveforms were attenuated: -4.1, -3.0, and 0 db for the sawtooth wave, the male and female EODs, respectively). Therefore, in an additional experiment the sine wave stimulus was attenuated to exactly match the weaker intensity of the fundamental frequency component of a sawtooth wave. All four fish tested still discriminated the rewarded sine wave from the unrewarded sawtooth wave in spite of their

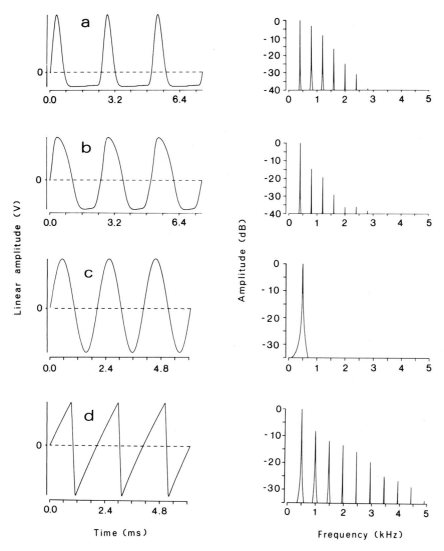

Fig. 4.66. Waveforms of electric signals (*left*) and Fourier amplitude spectra (*right*) as used for stimulation in conditioned discrimination experiments with *Eigenmannia lineata*. The *ordinates* of the *left* diagrams are arbitrary linear amplitudes (*V*), of the *right* diagrams amplitudes expressed as dB attenuation relative to the strongest spectral component of each waveform. *a* Electric organ discharges (EODs) of a male *Eigenmannia*, as generated by a microprocessor-based system for the digital synthesis of EODs (Kramer and Weymann 1987); *b* digitally synthesized EODs of a female *Eigenmannia*; *c* sine waves; *d* sawtooth waves generated by a function generator, all recorded from the water (Kramer and Zupanc 1986)

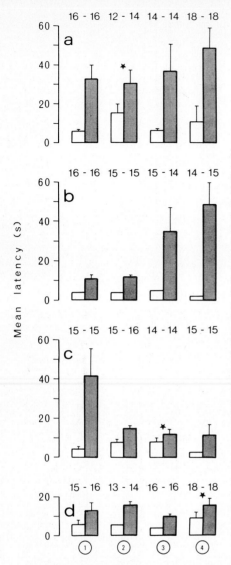

Fig. 4.67. Mean latencies of four isolated *Eigenmannia*, measured from the onset of an electric signal presented through a dipole to a fish's touching a feeder close to the dipole, in a food-rewarded conditioned discrimination experiment (modified T-maze). *Standard errors* are either shown or are too small to be drawn. Individual fish numbers are at the *bottom*. The number of test trials is indicated *above each column*. *Open columns* rewarded signals; *shaded* unrewarded signals. The differences among all paired columns are significant at $P<0.001$ (except * where $P<0.01$; one-tailed Mann-Whitney U-test). The rewarded and unrewarded signals, respectively, were *a* sine and sawtooth waves; *b* sine waves and male EODs; *c* female and male EODs; *d* sine and sawtooth waves of matched intensities of their fundamental frequency components, hence different peak-to-peak amplitudes (unlike the other pairs of signals). Note that in each fish, its latency to approach a dipole presenting a rewarded signal was much shorter than when the dipole presented an unrewarded signal. Experiment *c* shows discrimination of female from male EODs both having harmonic content. Experiment *d* shows that the fish recognized categorical rather than intensity differences in the stimuli presented (Kramer and Zupanc 1986)

identical f_1 intensities ($P<0.01$ in each fish, Fig. 4.67d). This shows that the fish indeed recognized categorical rather than intensity differences in the stimuli tested.

Which features of the signals are those that enable *Eigenmannia* to discriminate between waveforms? According to one hypothesis, fish are sensitive for differences in the intervals between zero-crossings of the two half-waves of a signal cycle (Gottschalk 1981). T-receptors are known to be phase-sensitive, and follow the discharge frequency cycle by cycle (see Chap. 2; Fig. 4.53). Intervals between zero-crossings are strictly equal in a sine or sawtooth wave (Fig. 4.66c,d) but markedly unequal in, for example, the male EOD (alternating short-long pattern; Fig. 4.66a) while they are almost equal in the female

EOD (Fig. 4.66b). [The apparent sensitivity of the JAR for this "temporal feature" of the EOD, as presumed by Gottschalk (1981) for $\Delta\frac{1}{2}F$ stimuli, was explained by variation of f_2 intensities of these stimuli; (Kramer 1985a). As experimentally shown, the JAR is *not* sensitive to this "temporal feature" per se; see also previous section of this chapter.]

The synthetic male EOD might have been discriminated from sine waves by its distinctive pattern of zero-crossings intervals. However, fish also discriminated sine from sawtooth waves although intervals between zero crossings were identical in both signals. A difference in this feature is, therefore, not required for discrimination of waveforms.

Therefore, *Eigenmannia* probably noticed the differences of the signals in harmonic content, or perhaps in an as yet unknown waveform parameter. One of the sensory requirements for detecting at least the low-frequency part of a signal's harmonic content, of fundamental frequency within the range of the species' EOD frequency, is met according to Viancour (1979a): a set of differently tuned electroreceptors in the same fish (see also Fig. 4.59). *Eigenmannia's* frequency difference limen to sinusoids of up to 4 times its EOD frequency is below 0.2% of the stimulus frequency, one of the best values ever recorded in the whole animal kingdom (Kramer and Kaunzinger, in prep.).

Apart from the most economical and straightforward mode of signal analysis that *Eigenmannia* uses in the context of the JAR, sharp band-pass filtering (see previous section), the fish apparently possesses another, more refined sensory mechanism to test the fine detail of a stimulus wave. That improved sensory capacity is used in a more complex behavioral context than that of the JAR, and enables fish to discriminate between females and males. Apart from its performance, this putative sensory mechanism can, at present, only be defined negatively (for example, differences in the pattern of zero-crossings intervals among two stimulus waves are *not* required for successful discrimination if differences in harmonic component intensities are present). Conversely, intensity differences in spectral composition are not required if spectral components differ in phase relationship, as shown by the observation that *Eigenmannia* discriminates the two signals shown in Fig. 4.63 (Kramer and Otto, in prep.).

Spontaneous Preference Test for Sexually Dimorphic EODs. The differences between the sexually dimorphic EOD waveforms of *Eigenmannia* are detailed in Section 3.1.2.2 (Fig. 3.15; see also Fig. 4.66a,b). These differences were shown to be almost independent of environmental changes such as temperature and water conductivity (Sect. 3.1.2.2, Fig. 3.17); they thus reliably signal the sex and age of the sender.

Fish were tested for their spontaneous preference for the synthetic male versus female EOD, simultaneously presented through two dipoles which were symmetrically placed to the left and right of a fish's daytime hiding place. The frequency differences from the fish's EOD frequency were ± 35 Hz; all stimulus conditions were randomly permuted according to a randomized blocks design (Cochran and Cox 1957).

Of 11 fish exposed to the paired stimulation of synthetic female and male EODs, eight showed a statistically significant preference for female EODs

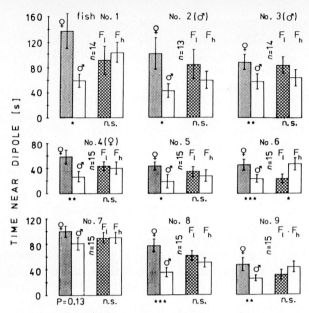

Fig. 4.68. Spontaneous preference of *Eigenmannia lineata* for synthetic *female* EODs, as compared to synthetic *male* EODs (*left pairs of columns* in each set), given as means ± SE, shown individually for nine fish. There was no significant effect of frequency that varied ±35 Hz from an experimental fish's EOD frequency (F_l for stimulus frequency 35 Hz *lower* than fish frequency, F_h for stimulus frequency 35 Hz *higher* than fish frequency; *right pairs of columns* in each set), except for fish No. 6 (which was attracted by "higher frequency" *and* "female synthetic EOD"). *Ordinate* the time (s) the fish spent within 20 cm of a dipole emitting synthetic male or female EODs. *NS* Difference nonsignificant; *, $P<0.05$; **, $P<0.02$; ***, $P<0.01$ (all two-tailed). *N* Number of days a complete random permutation of all stimulus conditions was run. *Fish No. 1* length 17.7 cm, P/N-ratio of its EOD = 0.57 (see Fig. 3.16), night observations; *fish No. 2* adult male 38 cm, P/N = 0.42, day; *fish No. 3* adult male 30 cm, P/N = 0.42, day; *fish No. 4* adult female gravid with eggs 15.5 cm, P/N = 0.60, day; *fish No. 5*, 15.5 cm, P/N = 0.72, night; *fish No. 6*, 19.6 cm, P/N = 0.75, night; *fish No. 7*, 13.5 cm, P/N not determined, night; *fish No. 8*, 14 cm, P/N = 0.82, night; *fish No. 9*, 15.8 cm, P/N = 0.70, day (Kramer and Otto 1988)

($P<0.05$, two-tailed; paired statistical tests), three fish had no significant preference (Fig. 4.68). All fish showed sidedness that was statistically balanced by the symmetrical experimental design.

Surprisingly, juvenile and adult fish of both sexes alike showed preference for female synthetic EODs. Frequency was irrelevant in all fish but one (which spent more time with a +35 Hz than with a −35 Hz synthetic EOD, relative to the fish's own frequency; it also showed a strong preference for the female EOD). In two independent studies on full-grown, gonadally ripe *Eigenmannia*, the EOD frequencies of females were higher than those of males (Hagedorn and Heiligenberg 1985; Kramer 1985a).

The adult males in particular showed signs of great excitement and vigorous behavior when in close contact with the active dipoles (for example, butting, rapidly swimming back and forth alongside a dipole, or rolling back and forth laterally over a dipole in parallel orientation). With the exception of the last behavior, similar behaviors have been observed by Hopkins (1974a) and Hagedorn and Heiligenberg (1985) during agonistic encounters and during artificial stimulation via electrical dipoles.

Eigenmannia species are gregarious fish (Lissmann 1961; B.Kramer, pers. field observ.); it therefore seems normal that experimentally isolated fish should join conspecifics (or electric dipoles simulating the presence of conspecifics). The preference of both females and juveniles for female synthetic EODs may be due to male aggressiveness (see Hagedorn and Heiligenberg 1985; Kramer 1985a).

The present results independently support the findings of a conditioned discrimination capacity of synthetic male and female EODs (Fig. 4.67). In the context of food-associated stimuli (Fig. 4.67) or social stimuli (Fig. 4.68) *Eigenmannia* seems to use a more complex type of signal analysis than it does in the context of the jamming avoidance response (JAR), sharp bandpass-filtering.

Although sufficient for the JAR, in the context of finding food and conspecifics the fish seem to need more information than amplitude and frequency difference of only a single sine wave component, the fundamental frequency f_1, of another fish's EOD. Frequency modulations (see Hopkins 1974a; Hagedorn and Heiligenberg 1985; Kramer 1987) are not needed for discrimination of male from female EODs. *Eigenmannia* is capable of sensing the fine detail of its sexually dimorphic EOD even when amplitude and frequency are not factors (Fig. 4.67); this information may direct its behavior without training in a biologically important manner, even outside a specifically reproductive context. The sensory mechanisms of that complex sensory capacity certainly deserve special attention.

Mate recognition by its sex-specific difference in fundamental frequency was suggested in the low-frequency (about 50 to 150 Hz at 25°C) wave fish *Sternopygus macrurus* (of the same family, Sternopygidae): males discharge at about one octave below females (see Sect. 4.1.2.2). As in *Eigenmannia*, the frequency difference is manifested only in the sexually mature, adult *Sternopygus* (see Sect. 3.2.2.2). *Sternopygus'* EOD has a very similar wave shape compared with that of *Eigenmannia's*; Gottschalk (1981) suggested a sexual dimorphism in EOD waveform resembling the one found in *Eigenmannia*, with males displaying lower P/N-ratios than females (P/N-ratio explained in Fig. 3.16). To date, we do not know the functional significance of this waveform variability in *Sternopygus*.

It will be fascinating to unravel the sensory mechanisms by which these highly specialized creatures analyze the signals they exchange for communication, and recognize conspecifics and mates. It would not be too surprising if this also meant significant progress for general physiology, with spinoff for the realm of audition which is so closely related phylogenetically to the electrical sense.

Conclusion

For good reviews of the significance of electroreception for general neurobiology and future prospects see Bullock and Szabo (1986) and Bullock (1986).

Most unusual for the behavioral scientist studying weakly electric fishes may be that by detecting electric organ discharges, a non-invasive technique, he records the activity of a brain nucleus; he may correlate this activity with overt behavior. He can also find the fish in the field by this kind of "natural neurophysiological telemetry". An immense advantage is that the electrical behavior of these fish is so simple it can be measured by a pre-programmed machine (that is, routinely and objectively).

An issue of the greatest importance is progress in geographical and microgeographical studies of variability that clarify species and subspecies differences as well as ranges, and apparent sibling species complexes. This work involves close collaboration with systematists. As huge river systems, for example, of the Amazon basin, have yet to be explored, there is great promise for new discoveries. Scientists of all disciplines need to be able to identify their study subject correctly so they can compare their results.

Waveforms of electric organ discharges are often complex and may, in certain species, be age- or sex-characteristic, or specific for different classes of individuals. Because of the physical nature of an electrical dipole field, the temporal waveform of voltage difference relative to a distant point does not change with angle or distance. The lack of any kind of transducer ahead of the dermal electroreceptors implies that a stimulus wave acts directly on the receptor cell, as generated at the source. (This is unlike the other octavolateral sensory cells with their mechanical transducers facing the stimulus; auditory, vestibular, lateral line). This situation offers unique opportunities to study the receptor or behavioral response to complex stimulus waves that are biologically relevant. This direction of research using natural or modified stimulus waveforms that are digitally synthesized, integrated with studies of the natural intra- and inter-specific variability of waveforms of electric organ discharges, has barely begun.

Another area that deserves more attention is the study of modulations of electric organ discharges correlated with overt behavior, and its experimental analysis in the widest sense. Only a few good examples exist; the richness of biological variation has barely been tapped.

Only by examining many different hypotheses and a variety of species during all the important activities of their lives may we hope to understand the sensory and central nervous basis of the complex behavioral patterns of these remarkable animals, their ecology, and evolution.

References

Albe-Fessard D, Chagas C (1954) Etude de la sommation à la jonction nerf-électroplaque chez le gymnote (*Electrophorus electricus*). J Physiol (Paris) 36:823–840

Altamirano M, Coates CW, Grundfest H, Nachmansohn D (1953) Mechanisms of bioelectric activity in electric tissue. I. The response to indirect and direct stimulation of electroplax of *Electrophorus electricus*. J Gen Physiol 37:91–110

Andres KH, Düring M von, Petrasch E (1988) The fine structure of ampullary and tuberous electroreceptors in the South American blind catfish *Pseudocetopsis spec*. Anat Embryol 177:523–535

Attneave F (1974) Informationstheorie in der Psychologie, 3rd edn. Hans Huber, Bern (Applications of information theory to psychology. Henry Holt, NY 1959)

Baerends GP, Brouwer R, Waterbolk HT (1955) Ethological studies on *Lebistes reticulatus*. I. An analysis of the male courtship pattern. Behaviour 8:249–335

Bässler G, Hilbig R, Rahmann H (1979) Untersuchungen zur circadianen Rhythmik der elektrischen und motorischen Aktivität von *Gnathonemus petersii* (Mormyridae, Pisces). Z Tierpsychol 49:156–163

Bass AH (1986) Electric organs revisited: evolution of a vertebrate communication and orientation organ. In: Bullock TH, Heiligenberg W (eds) Electroreception. Wiley, NY, pp 13–70

Bastian J (1977) Variations in the frequency response of electroreceptors dependent on receptor location in weakly electric fish (Gymnotoidei) with a pulse discharge. J Comp Physiol 121:53–64

Bastian J (1981) Electrolocation. I. How the electroreceptors of *Apteronotus albifrons* code for moving objects and other electrical stimuli. J Comp Physiol 144:465–479

Bastian J (1986) Electrolocation: behavior, anatomy and physiology. In: Bullock TH, Heiligenberg W (eds) Electroreception. Wiley, NY, pp 577–612

Bastian J (1987a) Electrolocation in the presence of jamming signals: behavior. J Comp Physiol 161:811–824

Bastian J (1987b) Electrolocation in the presence of jamming signals: Electroreceptor physiology. J Comp Physiol 161:825–836

Bauer R (1968) Untersuchungen zur Entladungstätigkeit und zum Beutefangverhalten des Zitterwelses *Malapterurus electricus* Gmelin 1789 (Siluroidea – Malapteruridae, Lacep. 1803). Z Vgl Physiol 59:371–402

Bauer R (1974) Electric organ discharge activity of resting and stimulated *Gnathonemus petersii* (Mormyridae). Behaviour 50:306–323

Bauer R (1979) Electric organ discharge (EOD) and prey capture behaviour in the electric eel, *Electrophorus electricus*. Behav Ecol Sociobiol 4:311–319

Bauer R, Kramer B (1974) Agonistic behaviour in mormyrid fish: latency relationship between the electric discharges of *Gnathonemus petersii* and *Mormyrus rume*. Experientia 30:51–52

Belbenoit P (1970) Conditionnement instrumental de l'électroperception des objets chez *Gnathonemus petersii* (Mormyridae, Teleostei, Pisces). Z Vgl Physiol 67:192–204

Belbenoit P (1972) Relations entre la motricité et la décharge électrique chez les Mormyridae (Teleostei). J Physiol (Paris) 65(2):197A

Belbenoit P, Bauer R (1972) Video recordings of prey capture behaviour and associated electric organ discharge of *Torpedo marmorata* (Chondrichthyes). Mar Biol 17:93–99

Belbenoit P, Moller P, Serrier J (1983) Do electric fish sleep? In: Koella WP (ed) Sleep 1982 (6th Eur Congr Sleep Res). Karger, Basel, pp 240–242
Belbenoit P, Moller P, Serrier J, Push S (1979) Ethological observations on the electric organ discharge behavior of the electric catfish, *Malapterurus electricus* (Pisces). Behav Ecol Sociobiol 4:321–330
Bell CC (1986) Electroreception in mormyrid fish. Central physiology. In: Bullock TH, Heiligenberg W (eds) Electroreception. Wiley, NY, pp 423–452
Bell CC (1989) Sensory coding and corollary discharge effects in mormyrid electric fish. J Exp Biol 146:229–253
Bell CC (1990) Mormyromast electroreceptor organs and their afferent fibers in mormyrid fish: III. Physiological differences between two morphological types of fibers. J Neurophysiol 63:319–332
Bell CC, Szabo T (1986) Electroreception in mormyrid fish. Central anatomy. In: Bullock TH, Heiligenberg W (eds) Electroreception. Wiley, NY, pp 375–421
Bell CC, Myers JP, Russell CJ (1974) Electric organ discharge patterns during dominance related behavior displays in *Gnathonemus petersii*. J Comp Physiol 92:201–228
Bell CC, Bradbury J, Russell CJ (1976) The electric organ of a mormyrid as a current and voltage source. J Comp Physiol 110:65–88
Bell CC, Libouban S, Szabo T (1983) Neural pathways related to the electric organ discharge command in mormyrid fish. J Comp Neurol 216:327–338
Bell CC, Zakon H, Finger TE (1989) Mormyromast electroreceptor organs and their afferent fibers in mormyrid fish: I. morphology. J Comp Neurol 286:391–407
Bénech V, Quensière J (1983) Migrations de poissons vers lac Tchad à la décrue de la plaine inondée du Nord–Cameroun. Rev Hydrobiol Trop 16:287–316
Bennett MVL (1967) Mechanisms of electroreception. In: Cahn P (ed) Lateral line detectors. Indiana Univ Press, Bloomington, pp 313–393
Bennett MVL (1971a) Electric organs. In: Hoar WS, Randall DJ (eds) Fish physiology, vol V. Academic Press, Lond NY, pp 347–491
Bennett MVL (1971b) Electroreception. In: Hoar WS, Randall DJ (eds) Fish physiology, vol V. Academic Press, Lond NY, pp 493–574
Bennett MVL, Obara S (1986) Ionic mechanisms and pharmacology of electroreceptors. In: Bullock TH, Heiligenberg W (eds) Electroreception. Wiley, NY, pp 157–181
Bennett MVL, Giménez M, Nakajima Y, Pappas GD (1964) Spinal and medullary nuclei controlling electric organ in the eel, *Electrophorus*. Biol Bull 127:362
Birkholz J (1969) Zufällige Nachzucht bei *Petrocephalus bovei*. Aquarium 3:201–203
Black-Cleworth P (1970) The role of electrical discharges in the non-reproductive social behaviour of *Gymnotus carapo* (Gymnotidae, Pisces). Anim Behav 3:1–77
Blake RW (1983a) Fish locomotion. Cambridge Univ Press, Cambridge
Blake RW (1983b) Swimming in the electric eels and knife fishes. Can J Zool 61:1432–1441
Blüm V (1986) Vertebrate reproduction. Springer, Berlin Heidelberg New York Tokyo
Bodznick D, Boord RL (1986) Electroreception in chondrichthyes. Central anatomy and physiology. In: Bullock TH, Heiligenberg W (eds) Electroreception. Wiley, NY, pp 225–256
Bodznick D, Northcutt RG (1981) Electroreception in lampreys: Evidence that the earliest vertebrates were electroreceptive. Science 212:465–467
Bonn U, Kramer B (1987) Distribution of monoamine-containing neurons in the brain of the weakly electric teleost, *Eigenmannia lineata* (Gymnotiformes: Rhamphichthyidae). Z Mikrosk Anat Forsch 101:339–362
Boulenger GA (1909) Catalogue of the fresh-water fishes of Africa in the British Museum (Natural History), vol 1. British Museum (NH), London
Box GEP, Jenkins GM (1976) Time series analysis: forecasting and control. Holden-Day, San Francisco
Braford MR Jr (1986) African Knifefishes. The Xenomystines. In: Bullock TH, Heiligenberg W (eds) Electroreception. Wiley, NY, pp 453–464

Braitenberg V (1977) On the texture of brains. Springer, Berlin Heidelberg New York, pp 64–81

Bratton BO (1987) Signalling with the electric organ discharge in the teleost fish *Pollimyrus isidori* before and during reproductive behavior. PhD Thesis, Univ Regensburg

Bratton BO, Ayers LJ (1987) Observations on the electric organ discharge of two skate species (Chondrichthyes: Rajidae) and its relationship to behaviour. Environ Biol Fishes 20:241–254

Bratton BO, Kramer B (1988) Intraspecific variability of the pulse-type discharges of the African electric fishes, *Pollimyrus isidori* and *Petrocephalus bovei* (Mormyridae, Teleostei), and their dependence on water conductivity. Exp Biol 47:227–238

Bratton BO, Kramer B (1989) Patterns of the electric organ discharge during courtship and spawning in the mormyrid *Pollimyrus isidori*. Behav Ecol Sociobiol 24:349–368

Bruns V (1971) Elektrisches Organ von *Gnathonemus* (Mormyridae). Z Zellforsch 122:538–563

Bullock TH (1969) Species differences in effect of electroreceptor input on electric organ pacemakers and other aspects of behavior in electric fish. Brain Behav Evol 2:85–118

Bullock TH (1970) The reliability of neurons. J Gen Physiol 55:565–584

Bullock TH (1974) An essay on the discovery of sensory receptors and the assignment of their functions together with an introduction to electroreceptors. In: Fessard A (ed) Handbook of sensory physiology, vol III/3. Springer, Berlin Heidelberg New York, pp 1–12

Bullock TH (1982) Electroreception. Annu Rev Neurosci 5:121–170

Bullock TH (1986) Significance of findings on electroreception for general neurobiology. In: Bullock TH, Heiligenberg W (eds) Electroreception. Wiley, NY, pp 651–674

Bullock TH, Heiligenberg W (eds) (1986) Electroreception. Wiley, NY

Bullock TH, Szabo T (1986) Introduction. In: Bullock TH, Heiligenberg W (eds) Electroreception. Wiley, NY, pp 1–12

Bullock TH, Hamstra RH, Scheich H (1972a) The jamming avoidance response of high frequency electric fish. I. General features. J Comp Physiol 77:1–22

Bullock TH, Hamstra RH, Scheich H (1972b) The jamming avoidance response of high frequency electric fish. II. Quantitative aspects. J Comp Physiol 77:23–48

Bullock TH, Behrend K, Heiligenberg W (1975) Comparison of the jamming avoidance response in gymnotoid and gymnarchid electric fish: a case of convergent evolution of behavior and its sensory basis. J Comp Physiol 103:97–121

Burkhardt D, de la Motte I (1985) Selective pressures, variability, and sexual dimorphism in stalk-eyed flies (Diopsidae). Naturwissenschaften 72:204–206

Burkhardt D, de la Motte I (1987) Physiological, behavioural, and morphometric data elucidate the evolutive significance of stalked eyes in Diopsidae (Diptera). Entomol Gen 12:221–233

Burkhardt D, de la Motte I (1988) Big 'antlers' are favoured: female choice in stalk-eyed flies (Diptera, Insecta), field collected harems and laboratory experiments. J Comp Physiol 162:649–652

Carr CE, Maler L (1986) Electroreception in gymnotiform fish. Central anatomy and physiology. In: Bullock TH, Heiligenberg W (eds) Electroreception. Wiley, NY, pp 319–373

Coates CW, Altamirano M, Grundfest H (1954) Activity in electrogenic organs of knifefishes. Science 120:845–846

Cochran WG, Cox GM (1957) Experimental designs, 2nd edn. Wiley, NY

Colinvaux, PA (1989) The past and future Amazon. Sci Am 260:68–74

Cox RT (1938) The electric eel at home. NY Zool Soc 41:59–65

Crawford JD, Hagedorn M, Hopkins CD (1986) Acoustic communication in an electric fish, *Pollimyrus isidori* (Mormyridae). J Comp Physiol 159:297–310

Daget J, Durand JR (1981) Poissons. In: Durand JR, Lévêque C (eds) Flore et faune aquatiques de l'Afrique sahélo-soudanienne, vol 2. Office Rech Sci Tech Outre-Mer (ORSTOM), Paris, pp 687–771

Daget J, Gosse JP, Thys van den Audenaerde DFE (eds) (1984) Check-list of the freshwater fishes of Africa (CLOFFA), vol 1. Office Rech Sci Tech Outre-Mer (ORSTOM), Paris/Mus R Afr Cent (MRAC), Tervuren (Belgium)

Daget J, Gosse JP, Thys van den Audenaerde DFE (eds) (1986) Check-list of the freshwater fishes of Africa (CLOFFA), vol 3. Inst R Sci Nat Belgique (ISNB), Bruxelles/Mus R Afr Cent (MRAC), Tervuren (Belgium)/Office Rech Sci Tech Outre-Mer (ORSTOM), Paris

Darwin CR (1975; first published in 1859) On the origin of species by means of natural selection, or the perservation of favoured races in the struggle for life. Cambridge University Press, New York

Davis EA, Hopkins CD (1987) Behavioural analysis of electric signal localization in the electric fish, *Gymnotus carapo* (Gymnotiformes). Anim Behav 36:1658–1671

Denizot JP, Kirschbaum F, Westby GWM, Tsuji S (1978) The larval electric organ of the weakly electric fish *Pollimyrus (Marcusenius) isidori* (Mormyridae, Teleostei). J Neurocytol 7:165–181

Dewsbury (1966) Stimulus-produced changes in the discharge rate of an electric fish and their relation to arousal. Psychol Rec 16:495–504

Dijkgraaf S, Verheijen FJ (1950) Neue Versuche über das Tonunterscheidungsvermögen bei der Elritze. Z Vgl Physiol 32:248–256

Durand JR, Lévêque C (eds) (1981) Flore et faune aquatiques de l'Afrique sahélo-soudanienne (2 vols). Office Rech Sci Tech Outre-Mer (ORSTOM), Paris

Dye JC, Meyer JH (1986) Central control of the electric organ discharge in weakly electric fish. In: Bullock TH, Heiligenberg W (eds) Electroreception. Wiley, NY, pp 71–102

Eckert R, Randall D, Augustine G (1988) Animal physiology. Mechanisms and adaptations, 3rd edn. Freeman, San Francisco

Elekes K, Szabo T (1985) The mormyrid brain stem. III. Ultrastructure and synaptic organization of the medullary "pacemaker" nucleus. Neuroscience 15:457–492

Elekes K, Ravaille M, Bell CC, Libouban S, Szabo T (1985) The mormyrid brainstem. II. The medullary electromotor relay nucleus. An ultrastructural HRP study. Neuroscience 15:417–430

Elepfandt A (1985) Wave frequency recognition and absolute pitch for water waves in the clawed frog, *Xenopus laevis*. J Comp Physiol 158:235–238

Ellis MM (1913) The gymnotid eels of tropical America. Mem Carnegie Mus (Pittsburgh) 6:109–195

Endler J (1977) Geographic variation, speciation and clines. Princeton Univ Press, Princeton

Enger PS, Szabo T (1968) Effect of temperature on the discharge rates of the electric organ of some gymnotoids. Comp Biochem Physiol 27:625–627

Erskine FT, Howe DW, Weed BC (1966) The discharge period of the weakly electric fish *Sternarchus albifrons*. Am Zool 6:521

Farabaugh SM (1982) The ecological and social significance of duetting. In: Kroodsma DE, Miller EH (eds) Acoustic communication in birds, vol 2. Academic Press, Lond NY, p 85–124

Fay RR (1970) Auditory frequency discrimination in the goldfish (*Carassius auratus*). J Comp Physiol Psychol 73:175–180

Fay RR (1974) Frequency discrimination in vertebrates. J Acoust Soc Am 56:206–209

Fay RR (1982) Neural mechanisms of an auditory temporal discrimination by the goldfish. J Comp Physiol 147:201–216

Fay RR, Passow B (1982) Temporal discrimination in the goldfish. J Acoust Soc Am 72:753–760

Feng AS (1977) The role of the electrosensory system in postural control in the weakly electric fish *Eigenmannia virescens*. J Neurobiol 8:429–437

Fessard A (1958) Les organes électriques. In: Grassé PP (ed) Traité de Zoologie, Anatomie, Systématique, Biologie, vol XIII. Masson, Paris, pp 1143–1238

Fessard A (ed) (1974) Electroreceptors and other specialized receptors in lower vertebrates. Handbook of sensory physiology, vol III/3. Springer, Berlin Heidelberg New York

Fiedler K (1967) Ethologische und neuroanatomische Auswirkungen von Vorderhirnexstirpationen bei Meerbrassen (*Diplodus*) und Lippfischen (*Crenilabrus*, Perciformes, Teleostei). J Hirnforsch 9:481–563

Finger TE, Bell CC, Carr CE (1986) Comparisons among electroreceptive teleosts: why are electrosensory systems so similar? In: Bullock TH, Heiligenberg W (eds) Electroreception. Wiley, NY, pp 465–481

Fink SV, Fink WL (1981) Interrelationships of the ostariophysan fishes (Teleostei). Zool J Linn Soc 72:297–353

Fittkau EJ (1985) Ökologische und faunenhistorische Zoogeographie der tropischen Regenwälder – Versuch eines Vergleichs (Ecological and historical zoogeography of tropical rain forests: attempt of a comparison) Verh Dtsch Zool Ges 78:137–146

Fraenkel GS, Gunn DL (1940) The orientation of animals. Oxford Univ Press, Oxford (2nd edn. Dover Publ Inc, NY 1961)

Fritzsch B, Münz H (1986) Electroreception in amphibians. In: Bullock TH, Heiligenberg W (eds) Electroreception. Wiley, NY, pp 483–496

Geller W (1984) A toxicity warning monitor using the weakly electric fish, *Gnathonemus petersii*. Water Res 18:1285–1290

Géry J (1969) The freshwater fishes of South America. In: Fittkau EJ, Illies J, Klinge H et al. (eds) Biogeography and ecology in South America. Monographiae Biologicae no 9. Junk Publ, The Hague, pp 828–848

Gosse JP (1963) Le milieu aquatique et l'écologie des poissons dans la région de Yangambi. Ann Mus R Afr Cent (Tervuren, Belgique) Sér Sci Zool 116:113–271

Gosse JP (1984) Mormyriformes. In: Daget J, Gosse JP, Thys van den Audenaerde DFE (eds) Check-list of the freshwater fishes of Africa. Office Rech Sci Tech Outre-Mer (ORSTOM), Bondy (France)/Mus R Afr Cent (MRAC), Tervuren (Belgium), pp 63–124

Gottschalk B (1981) Electrocommunication in gymnotid wave fish: significance of a temporal feature in the electric organ discharge. In: Szabo T, Czeh G (eds) Sensory physiology of aquatic lower vertebrates. Adv Physiol Sci, Pergamon Press / Akademiai Kiado, Budapest, 31:255–277

Gottschalk B, Scheich H (1979) Phase sensitivity and phase coupling: common mechanisms for communication behaviors in gymnotid wave and pulse species. Behav Ecol Sociobiol 4:395–408

Goulding M (1980) The fishes and the forest. Explorations in Amazonian natural history. Univ California Press, Berkeley 1980

Graff C (1986) Signaux électriques et comportement social du poisson à faibles décharges, *Marcusenius macrolepidotus* (Mormyridae, Teleostei). PhD Thesis, Univ Paris-Sud Centre d'Orsay

Grant K, Bell CC, Clausse S (1986) Morphology and physiology of the brainstem nuclei controlling the electric organ discharge in mormyrid fish. J Comp Neurol 245:514–530

Griffin DR (1958) Listening in the dark. Yale Univ Press, Princeton

Grundfest H (1957) The mechanisms of discharge of the electric organ in relation to general and comparative electrophysiology. Progr Biophys Chem 7:3–85

Hagedorn M (1986) The ecology, courtship, and mating of gymnotiform electric fish. In: Bullock TH, Heiligenberg W (eds) Electroreception. Wiley, NY, pp 497–525

Hagedorn M (1988) Ecology and behavior of a pulse-type electric fish, *Hypopomus occidentalis* (Gymnotiformes, Hypopomidae), in a fresh water stream in Panama. Copeia 1988(2):324–335

Hagedorn M, Carr CE (1985) Single electrocytes produce a sexually dimorphic signal in South American electric fish, *Hypopomus occidentalis* (Gymnotiformes, Hypopomidae). J Comp Physiol 156:511–523

Hagedorn M, Finger T (1986) Two species of *Synodontis* are weakly electric. Abstr 1st Int Congr Neuroethol, Tokyo, p 98

Hagedorn M, Heiligenberg W (1985) Court and spark: electric signals in the courtship and mating of gymnotoid fish. Anim Behav 33:254–265

Hagedorn M, Zelick R (1989) Relative dominance among males is expressed in the electric organ discharge characteristics of a weakly electric fish. Anim Behav 38:520–525

Harder W (1968) Die Beziehungen zwischen Elektrorezeptoren, elektrischem Organ, Seitenlinienorganen und Nervensystem bei den Mormyridae (Teleostei, Pisces). Z Vgl Physiol 59:272–318

Harder W, Uhlemann H (1967) Zum Frequenzverhalten von *Gymnarchus niloticus* Cuv. (Mormyriformes, Teleostei). Z Vgl Physiol 54:85–88

Harder W, Schief A, Uhlemann H (1964) Zur Funktion des elektrischen Organs von *Gnathonemus petersii* (Gthr. 1862). Z Vgl Physiol 48:302–311

Harder W, Schief A, Uhlemann H (1967) Zur Empfindlichkeit des schwachelektrischen Fisches *Gnathonemus petersii* (Gthr. 1862) (Mormyriformes, Teleostei) gegenüber elektrischen Feldern. Z Vgl Physiol 54:89–108

Heiligenberg W (1974) Electrolocation and jamming avoidance in a *Hypopygus* (Rhamphichthyidae, Gymnotoidei), an electric fish with pulse-type discharge. J Comp Physiol 91:223–240

Heiligenberg W (1976) Electrolocation and jamming avoidance in the mormyrid fish *Brienomyrus*. J Comp Physiol 109:357–372

Heiligenberg W (1977) Principles of electrolocation and jamming avoidance in electric fish. Studies of brain function, vol 1. Springer, Berlin Heidelberg New York, pp 1–85

Heiligenberg W (1986) Jamming avoidance responses: model systems for neuroethology. In: Bullock TH, Heiligenberg W (eds) Electroreception. Wiley, NY, pp 613–649

Heiligenberg W (1988) The neuronal basis of electrosensory perception and its control of a behavioral response in a weakly electric fish. In: Atema J, Fay RR, Popper AN, Tavolga WN (eds) Sensory biology of aquatic animals. Springer, Berlin Heidelberg New York Tokyo, pp 851–868

Heiligenberg W, Bastian J (1980) Species specificity of electric organ discharges in sympatric gymnotoid fish of the Rio Negro. Acta Biol Venez 10:187–203

Heiligenberg W, Bastian J (1984) The electric sense of weakly electric fish. Annu Rev Physiol 46:561–583

Heiligenberg W, Baker C, Bastian J (1978a) The jamming avoidance response in gymnotid pulse species: a mechanism to minimize the probability of pulse-train coincidence. J Comp Physiol 124:211–224

Heiligenberg W, Baker C, Matsubara J (1978b) The jamming avoidance response in *Eigenmannia* revisited: the structure of a neural democracy. J Comp Physiol 127:267–286

Heinrich U (1987) Das Unterscheidungsvermögen des Nilhechtes *Pollimyrus isidori* gegenüber freilaufenden Pulsraten. Diploma Thesis, Univ Regensburg (FRG)

Hennig W (1966) Phylogenetic systematics. Univ Illinois Press, Urbana

Heymer A, Harder W (1975) Erstes Auftreten der elektrischen Entladungen bei einem jungen Mormyriden. Naturwissenschaften 62:489

Hildebrand M (1982) Analysis of vertebrate structure, 2nd edn. Wiley, NY

Hoedeman JJ (1962a) Notes on the ichthyology of Surinam and other Guianas. 9. New records of gymnotid fishes. Bull Aquat Biol 3(26):53–60

Hoedeman JJ (1962b) Notes on the ichthyology of Surinam and other Guianas. 11. New Gymnotoid fishes from Surinam and French Guiana, with additional records and a key to the groups and species from Guiana. Bull Aquat Biol 3(30):97–108

Hopkins CD (1972) Sex differences in electric signaling in an electric fish. Science 176:1035–1037

Hopkins CD (1974a) Electric communication: functions in the social behavior of *Eigenmannia virescens*. Behaviour 50:270–305

Hopkins CD (1974b) Electric communication in the reproductive behavior of *Sternopygus macrurus* (Gymnotoidei). Z Tierpsychol 35:518–535
Hopkins CD (1974c) Electric communication in fish. Am Sci 62:426–437
Hopkins CD (1976) Stimulus filtering and electroreception: tuberous electroreceptors in three species of gymnotoid fish. J Comp Physiol 111:171–207
Hopkins CD (1983) Functions and mechanisms in electroreception. In: Northcutt RG, Davis RE (eds) Fish neurobiology. Univ Michigan Press, Ann Arbor, pp 215–259
Hopkins CD (1986a) Behavior of mormyridae. In: Bullock TH, Heiligenberg W (eds) Electroreception. Wiley, NY, pp 527–576
Hopkins CD (1986b) Temporal structure of non-propagated electric communication signals. Brain Behav Evol 28:43–59
Hopkins CD, Bass AH (1981) Temporal coding of species recognition signals in an electric fish. Science 212:85–87
Hopkins CD, Heiligenberg W (1978) Evolutionary designs for electric signals and electroreceptors in gymnotoid fishes of Surinam. Behav Ecol Sociobiol 3:113–134
Hopkins CD, Westby GWM (1986) Time domain processing of electric organ discharge waveforms by pulse-type electric fish. Brain Behav Evol 29:77–104
Hoshimiya N, Shogen K, Matsuo, T, Chichibu S (1980) The *Apteronotus* EOD field: waveform and EOD field simulation. J Comp Physiol 135:283–290
Hudspeth AJ (1985) The cellular basis of hearing: the biophysics of hair cells. Science 230:745–752
Hudspeth AJ, Corey DP (1977) Sensitivity, polarity and conductance change in the response pattern of vertebrate hair cells to controlled mechanical stimuli. Proc Natl Acad Sci 74:2407–2411
Jacobs DW, Tavolga WN (1968) Acoustic frequency discrimination in the goldfish. Anim Behav 16:67–71
Jäger U (1974) Geruchsrezeption und Entladungsaktivität bei dem schwachelektrischen Fisch *Gnathonemus petersii* (Günther 1862) (Mormyridae, Teleostei). PhD Thesis, Univ Saarbrücken (FRG)
Julien RM (1988) A primer of drug action, 5th edn. Freeman, San Francisco
Kalmijn AJ (1974) The detection of electric fields from inanimate and animate sources other than electric organs. In: Fessard A (ed) Handbook of sensory physiology, vol III/3. Springer, Berlin Heidelberg New York, pp 147–200
Kalmijn AJ (1988) Detection of weak electric fields. In: Atema J, Fay RR, Popper AN, Tavolga WN (eds) Sensory biology of aquatic animals. Springer, Berlin Heidelberg New York Tokyo, pp 151–186
Keeton WT (1979) Avian orientation and navigation. Annu Rev Physiol 41:353–366
Keeton WT (1980) Biological science, 3rd edn. Norton, NY Lond
Kellaway P (1946) The part played by electric fish in the early history of bioelectricity and electrotherapy. Bull Hist Med 20:112–137
Keynes RD, Martins-Ferreira H (1953) Membrane potentials in the electroplates of the electric eel. J Physiol 119:315–351
Kirschbaum F (1975) Environmental factors control the periodical reproduction of tropical electric fish. Experientia 31:1159–1160
Kirschbaum F (1977) Electric organ ontogeny. Distinct larval organ precedes the adult organ in weakly electric fish. Naturwissenschaften 64:387–388
Kirschbaum F (1981) Ontogeny of both larval electric organ and electromotoneurones in *Pollimyrus isidori* (Mormyridae, Teleostei). In: Szabo T, Czeh G (eds) Sensory physiology of aquatic lower vertebrates. Adv Physiol Sci, Pergamon Press / Akademiai Kiado, Budapest, 31:129–157
Kirschbaum F (1983) Myogenic electric organ precedes the neurogenic organ in apteronotid fish. Naturwissenschaften 70:205–207
Kirschbaum F (1984) Reproduction of weakly electric teleosts: just another example of convergent development? Environ Biol Fishes 10:3–14
Kirschbaum F (1987) Reproduction and development of the weakly electric fish, *Pollimyrus isidori* (Mormyridae, Teleostei) in captivity. Environ Biol Fishes 20:11–31

Kirschbaum F, Westby GWM (1975) Development of the electric discharge in mormyrid and gymnotid fish (*Marcusenius* sp. and *Eigenmannia virescens*). Experientia 31:1290–1293

Kirschvink JL, Jones DS, MacFadden BJ (1985) (eds) Magnetite biomineralization and magnetoreception in organisms. Plenum, New York

Knöppel HA (1970) Food of Central Amazonian Fishes. Contribution to the nutrient-ecology of Amazonian rain-forest streams. Amazoniana 2:257–352

Knudsen EI (1974) Behavioral thresholds to electric signals in high frequency electric fish. J Comp Physiol 91:333–353

Knudsen EI (1975) Spatial aspects of the electric fields generated by weakly electric fish. J Comp Physiol 99:103–118

Kramer B (1971) Zur hormonalen Steuerung von Verhaltensweisen der Fortpflanzung beim Sonnenbarsch *Lepomis gibbosus* (L.) (Centrarchidae, Teleostei). Z Tierpsychol 28:351–386

Kramer B (1973) Chemische Wirkstoffe im Nestbau-, Sexual- und Kampfverhalten des Sonnenbarsches *Lepomis gibbosus* (L.) (Centrarchidae, Teleostei). Z Tierpsychol 32:353–373

Kramer B (1974) Electric organ discharge interaction during interspecific agonistic behaviour in freely swimming mormyrid fish. A method to evaluate two (or more) simultaneous time series of events with a digital analyser. J Comp Physiol 93:203–235

Kramer B (1976a) The attack frequency of *Gnathonemus petersii* towards electrically silent (denervated) and intact conspecifics, and towards another mormyrid (*Brienomyrus niger*). Behav Ecol Sociobiol 1:425–446

Kramer B (1976b) Flight-associated discharge pattern in a weakly electric fish, *Gnathonemus petersii* (Mormyridae, Teleostei). Behaviour 59:88–95

Kramer B (1976c) Electric signalling during aggressive behaviour in *Mormyrus rume* (Mormyridae, Teleostei). Naturwissenschaften 63:48

Kramer B (1978) Spontaneous discharge rhythms and social signalling in the weakly electric fish *Pollimyrus isidori* (Cuvier et Valenciennes) (Mormyridae, Teleostei). Behav Ecol Sociobiol 4:61–74

Kramer B (1979) Electric and motor responses of the weakly electric fish, *Gnathonemus petersii* (Mormyridae), to play-back of social signals. Behav Ecol Sociobiol 6:67–79

Kramer B (1984) Tranquillizer reduces electric organ discharge frequency in a teleost fish. Naturwissenschaften 71:99

Kramer B (1985a) Jamming avoidance in the electric fish *Eigenmannia*: harmonic analysis of sexually dimorphic waves. J Exp Biol 119:41–69

Kramer B (1985b) Kommunikation mit elektrischen Signalen bei Fischen. In: Franck D (ed) Verhaltensbiologie, 2nd edn. Deutscher Taschenbuch-Verlag, München/ Thieme, Stuttgart, pp 273–277

Kramer B (1987) The sexually dimorphic jamming avoidance response in the electric fish *Eigenmannia* (Teleostei, Gymnotiformes). J Exp Biol 130:39–62

Kramer B (1988) Schwachelektrische Fische: Ausweichreaktionen auf Störsender. Prax Naturwiss Biol 37(7):1–10

Kramer B (1990) Sexual signals in electric fish. Trends in Ecology and Evolution (TREE) 5:247–250

Kramer B, Bauer R (1976) Agonistic behaviour and electric signalling in a mormyrid fish, *Gnathonemus petersii*. Behav Ecol Sociobiol 1:45–61

Kramer B, Otto B (1988) Female discharges are more electrifying: spontaneous preference in the electric fish, *Eigenmannia*. Behav Ecol Sociobiol 23:55–60

Kramer B, Westby GWM (1985) No sex difference in the waveform of the pulse type electric fish, *Gnathonemus petersii* (Mormyridae). Experientia 41:1530–1531

Kramer B, Weymann D (1987) A microprocessor system for the digital synthesis of pulsed or continuous discharges of electric fish (or animal vocalizations). Behav Brain Res 23:167–174

Kramer B, Zupanc GKH (1986) Conditioned discrimination of electric waves differing only in form and harmonic content in the electric fish, *Eigenmannia*. Naturwissenschaften 73:679–680

Kramer B, Kirschbaum F, Markl H (1981a) Species specificity of electric organ discharges in a sympatric group of gymnotoid fish from Manaus (Amazonas). In: Szabo T, Czeh G (eds) Sensory physiology of aquatic lower vertebrates. Adv Physiol Sci, Pergamon Press / Akademiai Kiado, Budapest, 31:195–219

Kramer B, Tautz J, Markl H (1981b) The EOD sound response in weakly electric fish. J Comp Physiol 143:435–441

Kramer-Feil U (1976) Analyse der sinnesphysiologischen Eigenschaften der Einheit Elektrorezeptor (Mormyromast)-sensible Faser des schwachelektrischen Fisches *Gnathonemus petersii* (Mormyridae, Teleostei). PhD Thesis, Univ Frankfurt/Main (FRG)

Krebs JR, Davies NB (1987) An introduction to behavioural ecology (2nd edn). Blackwell's, Oxford

Kunze P (1989) Verhaltensänderungen durch Atrazin bei einem elektrischen Fisch (Change of behaviour in an electric fish). Z Wasser- Abwasser-Forsch 22:108–111

Kunze P, Wetzstein HU (1988) Apomorphine and haloperidol influence electric behaviour of a mormyrid fish. Z Naturforsch 43c:105–107

Lamarque P (1979) Le gymnote *Electrophorus electricus* et la pêche à l'électricité. La Pisciculture Française 15:22–26

Landsman RE, Moller P (1988) Testosterone changes the electric organ discharge and external morphology of the mormyrid fish, *Gnathonemus petersii* (Mormyriformes). Experientia 44:900–903

Langner G, Scheich H (1978) Active phase coupling in electric fish: behavioral control with microsecond precision. J Comp Physiol 128:235–240

Larimer JL, MacDonald JA (1968) Sensory feedback from electroreceptors to electromotor pacemaker centers in gymnotids. Am J Physiol 214:1253–1261

Lauder GV, Liem KF (1983) Patterns of diversity and evolution in ray-finned fishes. In: Northcutt RG, Davis RE (eds) Fish neurobiology. Univ Michigan Press, Ann Arbor, pp 1–24

Lévêque C, Paugy D (1984) Guide des poissons d'eau douce de la zone du programme de lutte contre l'onchocercose en Afrique de l'Ouest. Convention ORSTOM-OMS (ORSTOM, Office de la Recherche Scientifique et Technique Outre-Mer, Paris)

Libouban S, Szabo T, Ellis D (1981) Comparative study of the medullary command (pacemaker) nucleus in species of the four weakly electric fish families. In: Szabo T, Czeh G (eds) Sensory physiology of aquatic lower vertebrates. Adv Physiol Sci, Pergamon Press / Akademiai Kiado, Budapest, 31:95–106

Lissmann HW (1951) Continuous electrical signals from the tail of a fish, *Gymnarchus niloticus* Cuv. Nature (Lond) 167:201–202

Lissmann HW (1958) On the function and evolution of electric organs in fish. J Exp Biol 35:156–191

Lissmann HW (1961) Ecological studies on gymnotids. In: Chagas C, Paes de Carvalho A (eds) Bioelectrogenesis. Elsevier, Amst, pp 215–226

Lissmann HW (1963) Electric location by fishes. Sci Am, March 1963, pp 206–215

Lissmann HW, Machin KE (1958) The mechanism of object location in *Gymnarchus niloticus* and similar fish. J Exp Biol 35:451–486

Lissmann HW, Schwassmann HO (1965) Activity rhythm of an electric fish, *Gymnorhamphichthys hypostomus* Ellis. Z Vgl Physiol 51:153–171

Lowe-McConnell RH (1987) Ecological studies in tropical fish communities. Cambridge Univ Press, Cambridge

Lücker H (1982) Untersuchungen zur intraspezifischen Elektrokommunikation mittels der Latenzbeziehungen und zur interspezifischen Elektrokommunikation mittels der art- und aktivitätsspezifischen Entladungsmuster bei *Pollimyrus isidori* (Cuv. & Val.) und *Petrocephalus bovei* (Cuv. & Val.). PhD Thesis, Univ Konstanz (FRG)

Lücker H (1983) Species-specific discharge rhythms in mormyrids as a mechanism for species identification. Verh Dtsch Zool Ges 76:195 (in German)

Lücker H, Kramer B (1981) Development of a sex difference in the preferred latency response in the weakly electric fish, *Pollimyrus isidori* (Cuvier et Valenciennes) (Mormyridae, Teleostei). Behav Ecol Sociobiol 9:103–109

Lundberg JG, Mago-Leccia F (1986) A review of *Rhabdolichops* (Gymnotiformes, Sternopygidae), a genus of South American freshwater fishes, with descriptions of four new species. Proc Acad Nat Sci Philadelphia 138(1):53–85

Lundberg JG, Stager JC (1985) Microgeographic diversity in the neotropical knife-fish *Eigenmannia macrops* (Gymnotiformes, Sternopygidae). Environ Biol Fishes 13:173–181

Lundberg JG, Lewis WM, Saunders III JF, Mago-Leccia F (1987) A major food web component in the Orinoco river channel: evidence for planktivorous electric fishes. Science 237:81–83

MacDonald JA, Larimer JL (1970) Phase-sensitivity of *Gymnotus carapo* to low-amplitude electrical stimuli. Z Vgl Physiol 70:322–334

Machemer H (1988) Galvanotaxis: Grundlagen der elektro-mechanischen Kopplung und Orientierung bei Paramecium. In: Zupanc GKH (ed) Praktische Verhaltensbiologie. Parey, Berlin, pp 58–82

Mago-Leccia F (1978) Los peces de la familia Sternopygidae de Venezuela, incluyendo un descripcion de la osteologia de *Eigenmannia virescens* y una nueva definicion y clasificacion del orden Gymnotiformes. Acta Cient Venez 29(Supl 1):1–89

Mago-Leccia F, Zaret TM (1978) The taxonomic status of *Rhabdolichops troscheli* (Kaup, 1856), and speculations on gymnotiform evolution. Environ Biol Fishes 3:379–384

Mago-Leccia F, Lundberg JG, Baskin JN (1985) Systematics of the South American freshwater fish genus *Adontosternarchus* (Gymnotiformes, Apteronotidea). Contr Sci Nat Hist Mus Los Angeles County No. 358:1–19

Mandriota FJ, Thompson RL, Bennett MVL (1965) Classical contitioning of electric organ discharge rate in mormyrids. Science 150:1740–1742

Manning A (1979) An introduction to animal behaviour, 3rd edn. Edward Arnold, London

Matsubara J, Heiligenberg W (1978) How well do electric fish electrolocate under jamming? J Comp Physiol 125:285–290

McCormick CA, Popper AN (1984) Auditory sensitivity and psychophysical tuning curves in the elephant nose fish, *Gnathonemus petersii*. J Comp Physiol 155:753–761

Meszler RM, Pappas GD, Bennett MVL (1974) Morphology of the electromotor system in the spinal cord of the electric eel, *Electrophorus electricus*. J Neurocytol 3:251–261

Meyer DL, Heiligenberg W, Bullock TH (1976) The ventral substrate response. A new postural control mechanism in fishes. J Comp Physiol 109:59–68

Meyer H (1982) Behavioral responses of weakly electric fish to complex impedances. J Comp Physiol 145:459–470

Meyer JH (1983) Steroid influences upon the discharge frequencies of a weakly electric fish. J Comp Physiol 153:29–37

Meyer JH (1984) Steroid influences upon discharge frequencies of intact and isolated pacemakers of weakly electric fish. J Comp Physiol 154:659–668

Möhres FP (1957) Elektrische Entladungen im Dienste der Revierabgrenzung bei Fischen. Naturwissenschaften 44:431–432

Möhres FP (1961) Die elektrischen Fische. Natur Volk (Senckenberg Mus, Frankfurt/Main, FRG) 91:1–13

Moller P (1969) Ein Beitrag zur Frage nach der Kommunikation unter schwach elektrischen Fischen (*Gnathonemus moori*, Mormyridae). Zool Anz (Suppl) 33:482–489

Moller P (1970) Communication in weakly electric fish, *Gnathonemus petersii* (Mormyridae). I. Variation of electric organ discharge frequency elicited by controlled electric stimuli. Anim Behav 18:768–786

Moller P (1976) Electric signals and schooling behavior in a weakly electric fish, *Marcusenius cyprinoides* (Mormyriformes). Science 193:697–699
Moller P (1980) Electroperception. Oceanus 23:44–54
Moller P, Bauer R (1973) 'Communication' in weakly electric fish, *Gnathonemus petersii* (Mormyridae) II. Interaction of electric organ discharge activities of two fish. Anim Behav 21:501–512
Moller P, Serrier J (1986) Species recognition in mormyrid weakly electric fish. Anim Behav 34:333–339
Moller P, Serrier J, Belbenoit P (1976) Electric organ discharges of the weakly electric fish *Gymnarchus niloticus* (Mormyriformes) in its natural habitat. Experientia 32:1007
Moller P, Serrier J, Belbenoit P, Push S (1979) Notes on ethology and ecology of the Swashi River Mormyrids (Lake Kainji, Nigeria). Behav Ecol Sociobiol 4:357–368
Murray RW (1960) Electrical sensitivity of the ampullae of Lorenzini. Nature (Lond) 187:957
Murray RW (1974) The ampullae of Lorenzini. In: Fessard A (ed) Handbook of sensory physiology, vol III/3. Springer, Berlin Heidelberg New York, pp 125–146
Nelson JS (1984) Fishes of the world, 2nd edn. Wiley, NY
Neuweiler G (1984) Foraging, echolocation and audition in bats. Naturwissenschaften 71:446–455
Northcutt RG (1986) Electroreception in nonteleost bony fishes. In: Bullock TH, Heiligenberg W (eds) Electroreception. Wiley, NY, pp 257–285
Parker GH, van Heusen AP (1917) The responses of the catfish, *Amiurus nebulosus*, to metallic and non-metallic rods. Am J Physiol 44:405–420
Pickens PE, McFarland WN (1964) Electric discharge and associated behaviour in the stargazer. Anim Behav 12:362–367
Paul D (1972) Zur Signalverarbeitung im elektrischen Empfangsorgan des schwachelektrischen Fisches *Gnathonemus petersii* (Mormyriformes, Teleostei). Z Vgl Physiol 76:193–203
Popp JW (1989) Methods of measuring avoidance of acoustic interference. Anim Behav 38:358–360
Popper AN, Fay RR (1973) Sound detection and processing by teleost fishes: a critical review. J Acoust Soc Am 53:1515–1529
Popper AN, Fay RR (1984) Sound detection and processing by teleost fish: a selective review. In: Bolis L, Keynes RD, Maddrell SHP (eds) Comparative physiology of sensory systems. Cambridge University Press, Cambridge, pp 67–101
Popper AN, Rogers PH, Saidel WM, Cox M (1988) Role of the fish ear in sound processing. In: Atema J, Fay RR, Popper AN, Tavolga WN (eds) Sensory biology of aquatic animals. Springer, Berlin Heidelberg New York Tokyo, pp 687–710
Purves WK, Orians GH (1987) Life. The science of biology, 2nd edn. Sinauer, Sunderland (Massachusetts)
Quastler, H (1958) A primer on information theory. In: Yockey HP, Platzman RL, Quastler H (eds) Symposium on information theory in biology. Pergamon Press, NY, pp 3–49
Rankin C, Moller P (1986) Social behavior of the African catfish, *Malapterurus electricus*, during intra- and interspecific encounters. Ethology 73:177–190
Raven PH, Johnson GB (1986) Biology. Times Mirror/Mosby College, St. Louis
Rigley L, Marshall JA (1973) Sound production by the elephant nose fish, *Gnathonemus petersii* (Pisces, Mormyridae). Copeia 1973(1):134–136
Roberts TR (1973) Ecology of fishes in the Amazon and Congo basins. In: Meggers BJ, Ayensu ES, Duckworth WD (eds) Tropical forest ecosystems in Africa and South America: a comparative review. Smithsonian Inst Press, Washington DC, pp 239–254
Roberts TR, Stewart DJ (1976) An ecological and systematic survey of fishes in the rapids of the lower Zaïre or Congo river. Bull Mus Comp Zool 147:239–317
Roederer JG (1975) Introduction to the physics and psychophysics of music, 2nd edn. Springer, Berlin Heidelberg New York

Ronan M (1986) Electroreception in cyclostomes. In: Bullock TH, Heiligenberg W (eds) Electroreception. Wiley, NY, pp 209–224

Rosen DE (1982) Teleostean interrelationships, morphological function and evolutionary inference. Am Zool 22:261–273

Roth A (1968) Electroreception in the catfish, *Amiurus nebulosus*. Z Vgl Physiol 61:196–202

Russell CJ, Myers JP, Bell CC (1974) The echo response in *Gnathonemus petersii* (Mormyridae). J Comp Physiol 92:181–200

Scheich H (1974) Neural analysis of waveform in the time domain: midbrain units in electric fish during social behavior. Science 185:365–367

Scheich H (1977a) Neural basis of communication in the high frequency electric fish, *Eigenmannia virescens* (jamming avoidance response). I. Open loop experiments and the time domain concept of signal analysis. J Comp Physiol 113:181–206

Scheich H (1977b) Neural basis of communication in the high frequency electric fish, *Eigenmannia virescens* (jamming avoidance response). II. Jammed electroreceptor neurons in the lateral line nerve. J Comp Physiol 113:207–227

Scheich H (1977c) Neural basis of communication in the high frequency electric fish, *Eigenmannia virescens* (jamming avoidance response). III. Central integration in the sensory pathway and control of the pacemaker. J Comp Physiol 113:229–255

Scheich H (1982) Biophysik der Elektrorezeption. In: Hoppe W, Lohmann W, Markl H, Ziegler H (eds) Biophysik, 2nd edn. Springer, Berlin Heidelberg New York, pp 791–805

Scheich H, Bullock TH (1974) The detection of electric fields from electric organs. In: Fessard A (ed) Handbook of sensory physiology, vol III/3. Springer, Berlin Heidelberg New York, pp 201–256

Scheich H, Ebbesson SOE (1983) Multimodal torus in the weakly electric fish *Eigenmannia*. Advances in Embryology and Cell Biology, vol 82. Springer, Berlin Heidelberg New York

Scheich H, Bullock TH, Hamstra RH Jr (1973) Coding properties of two classes of afferent nerve fibers: high frequency electroreceptors in the electric fish, *Eigenmannia*. J Neurophysiol 36:39–60

Scheich H, Gottschalk B, Nickel B (1977) The jamming avoidance response in *Rhamphichthys rostratus*: an alternative principle of time domain analysis in electric fish. Exp Brain Res 28:229–233

Schluger JH, Hopkins CD (1987) Electric fish approach stationary signal sources by following electric current lines. J Exp Biol 130:359–367

Schöne H (1980) Orientierung im Raum. Formen und Mechanismen der Lenkung des Verhaltens im Raum bei Tier und Mensch. Wiss Verlagsgesellschaft, Stuttgart

Schwassmann HO (1971a) Circadian activity patterns in gymnotid electric fish. In: Menaker M (ed) Biochronometry. Natl Acad Sci, Wash DC, pp 186–199

Schwassmann HO (1971b) Biological rhythms. In: Hoar WS, Randall DJ (eds) Fish physiology, vol 6. Academic Press, Lond NY, pp 371–428

Schwassmann HO (1976) Ecology and taxonomic status of different geographic populations of *Gymnorhamphichthys hypostomus* Ellis (Pisces, Cypriniformes, Gymnotoidei). Biotropica 8:25–40

Schwassmann HO (1978a) Activity rhythms in gymnotoid electric fishes. In: Thorpe JE (ed) Rhythmic activity of fishes. Academic Press, Lond NY, pp 235–241

Schwassmann HO (1978b) Ecological aspects of electroreception. In: Ali MA (ed) Sensory ecology. Plenum Press, NY, pp 521–533

Schwassmann HO (1978c) Times of annual spawning and reproductive strategies in Amazonian fishes. In: Thorpe JE (ed) Rhythmic activity of fishes. Academic Press, London, pp 187–200

Schwassmann HO (1984) Species of *Steatogenys* Boulenger (Pisces, Gymnotiformes, Hypopomidae). Bol Mus Para Emilio Goeldi Zool 1(1):97–114

Schwassmann HO, Carvalho ML (1985) *Archolaemus blax* Korringa (Pisces, Gymnotiformes, Sternopygidae) a redescription with notes on ecology. Spixiana 8(3):231–240

Serrier J (1973) Modifications instantanées du rythme de l'activité électrique d'un mormyre, *Gnathonemus petersii*, provoquées par la stimulation électrique artificielle de ses électrorécepteurs. J Physiol (Paris) 66:713–728

Serrier J (1974) Rythmes de décharges électrique chez des mormyres en conditions expérimentales. PhD Thesis, Univ Paris VI

Serrier J (1982) Comportement électrique des Mormyridae (Pisces). Électrogénèse en réponse à un signal exogène. Thesis (Docteur ès sciences naturelles), Univ Paris-Sud Centre d'Orsay

Serrier J, Moller P (1981) Social behavior in mormyrid fish (Mormyriformes, Pisces): short and long-term changes associated with repeated interactions. In: Szabo T, Czeh G (eds) Sensory physiology of aquatic lower vertebrates. Adv Physiol Sci, Pergamon Press / Akademiai Kiado, Budapest, 31:221–233

Shannon CE, Weaver W (1949) The mathematical theory of communication. Univ of Illinois Press, Urbana

Shumway CA, Zelick RD (1988) Sex recognition and neuronal coding of electric organ discharge waveform in the pulse-type weakly electric fish, *Hypopomus occidentalis*. J Comp Physiol 163:465–478

Siegel S (1956) Nonparametric statistics for the behavioral sciences. McGraw-Hill, NY

Sioli H (1983) Amazonien. Grundlagen der Ökologie des grössten tropischen Waldlandes. Wiss Verlagsgesellschaft, Stuttgart

Sioli H (ed) (1984) The Amazon. Limnology and landscape ecology of a mighty tropical river and its basin. Junk Publ, Dordrecht

Smith RJF (1985) The control of fish migration (Zoophysiology vol 17). Springer, Berlin Heidelberg New York Tokyo

Squire A, Moller P (1982) Effects of water conductivity on electrocommunication in the weak-electric fish *Brienomyrus niger* (Mormyriformes). Anim Behav 30:375–382

Srivastava CBL, Szabo T (1972) Development of electric organs of *Gymnarchus niloticus* (Fam. Gymnarchidae): I. Origin and histogenesis of electroplates. J Morphol 138:375–386

Srivastava CBL, Szabo T (1973) Development of electric organs of *Gymnarchus niloticus* (Fam. Gymnarchidae): II. Formation of spindle. J Morphol 140:461–466

Stanley SM (1979) Macroevolution: pattern and process. Freeman, San Francisco

Starck D (1978) Vergleichende Anatomie der Wirbeltiere auf evolutionsbiologischer Grundlage, vol 1. Springer, Berlin Heidelberg New York

Steinbach AB (1970) Diurnal movements and discharge characteristics of electric gymnotid fishes in the Rio Negro, Brazil. Biol Bull 138:200–210

Stetter H (1929) Untersuchungen über den Gehörsinn der Fische, besonders von *Phoxinus laevis* L. und *Amiurus nebulosus* Raf. Z Vgl Physiol 9:339–477

Stipetić E (1939) Über das Gehörorgan der Mormyriden. Z Vgl Physiol 26:740–752

Szabo (1958) Structure intime de l'organe électrique de trois mormyridés. Z Zellforsch 49:33–45

Szabo T (1960) Development of the electric organ of Mormyridae. Nature (Lond) 188:760–762

Szabo T (1961) Les organes électriques des mormyridés. In: Chagas C, Paes de Carvalho A (eds) Bioelectrogenesis. Elsevier, Amst, pp 20–24

Szabo T (1962a) The activity of cutaneous sensory organs in *Gymnarchus niloticus*. Life Sci 7:285–286

Szabo T (1962b) Organes sensoriels autoactifs dans le tégument de *Gymnarchus niloticus* (Poisson mormyroïde). Comptes Rendus Acad Sci 255:177–178

Szabo T (1965) Sense organs of the lateral line system in some electric fish of Gymnotidae, Mormyridae and Gymnarchidae. J. Morph. 117:229–250

Szabo T (1967) Activity of peripheral and central neurons involved in electroreception. In: Cahn P (ed) Lateral line detectors. Indiana University Press, Bloomington (Indiana), pp 295–311

Szabo T (1970a) Elektrische Organe und Elektrorezeption bei Fischen. Westdtsch Verlag, Rheinisch-Westfälische Akad Wiss, Vortr N 205:7–40

Szabo T (1970b) Über eine bisher unbekannte Funktion der sog. ampullären Organe bei *Gnathonemus petersii*. Z Vgl Physiol 66:164–175

Szabo T (1974) Anatomy of the specialized lateral line organs of electroreception. In: Fessard A (ed) Handbook of sensory physiology, vol III/3. Springer, Berlin Heidelberg New York, pp 13–58

Szabo T, Enger PS (1964) Pacemaker activity of the medullary nucleus controlling electric organs in high frequency gymnotid fish. Z Vgl Physiol 49:285–300

Szabo T, Fessard A (1974) Physiology of electroreceptors. In: Fessard A (ed) Handbook of sensory physiology, vol III/3. Springer, Berlin Heidelberg New York, pp 59–124

Szabo T, Moller P (1984) Neuroethological basis for electrocommunication. In: Bolis L, Keynes RD, Maddrell SHP (eds) Comparative physiology of sensory systems. Cambridge Univ Press, Cambridge, pp 455–474

Szabo T, Suckling EE (1964) L'arrêt occasionel de la décharge électrique continue du *Gymnarchus* est-il une réaction naturelle? Naturwissenschaften 51:92–94

Taverne L (1972) Ostéologie des genres *Mormyrus* Linné, *Mormyrops* Müller, *Hyperopisus* Gill, *Isichthys* Gill, *Myomyrus* Boulenger, *Stomatorhinus* Boulenger et *Gymnarchus* Cuvier. Considérations générales sur la systématique des poissons de l'ordre des Mormyriformes. Ann Mus R Afr Cent (Tervuren, Belgique) Sér Sci Zool 200:1–194

Tavolga WN (1971) Sound production and detection. In: Hoar WS, Randall DJ (eds) Fish physiology, vol 5. Academic Press, Lond New York, pp 135–205

Tavolga WN (1974) Signal/noise ratio and the critical band in fishes. J Acoust Soc Am 55:1323–1333

Teyssèdre C, Moller P (1982) The optomotor response in weak electric fish: can they see? Z Tierpsychol 60:306–312

Teyssèdre C, Boudinot M, Minisclou C (1987) Categorisation of interpulse intervals and stochastic analysis of discharge patterns in resting weak-electric fish (*Gnathonemus petersii*). Behaviour 102:264–282

Tinbergen N (1979) Instinktlehre. Parey, Berlin (The study of instinct. Oxford Univ Press, Lond 1951)

Toerring MJ, Belbenoit P (1979) Motor programme and electroreception in mormyrid fish. Behav Ecol Sociobiol 4:369–379

Toerring MJ, Moller P (1984) Locomotor and electric displays associated with electrolocation during exploratory behavior in mormyrid fish. Behav Brain Res 12:291–306

Toerring MJ, Serrier J (1978) Influence of water temperature on the electric organ discharge (EOD) of the weakly electric fish *Marcusenius cyprinoides* (Mormyridae). J Exp Biol 74:133–150

Thorpe WH (1972) Duetting and antiphonal song in birds: Its extent and significance. Behaviour (Suppl) 18:1–197

Valone JA (1970) Electrical emissions in *Gymnotus carapo* and their relation to social behaviour. Behaviour 37:1–14

van Oordt PGWJ (1987) Modern trends in reproductive endocrinology of teleosts. Proc 5th Congr Eur Ichthyol, Stockholm 1985, pp 247–268

Verworn M (1889) Psychophysiologische Protistenstudien. Experimentelle Untersuchungen. Gustav Fischer, Jena

Viancour TA (1979a) Electroreceptors of a weakly electric fish. I. Characterization of tuberous electroreceptor tuning. II. Individually tuned receptor oscillations. J Comp Physiol 133:317–338

Viancour TA (1979b) Peripheral electrosense physiology: a review of recent findings. J Physiol (Paris) 75:321–333

von der Emde G, Menne D (1989) Discrimination of insect wingbeat-frequencies by the bat *Rhinolophus ferrumequinum*. J Comp Physiol 164:663–671

von Frisch K (1938) The sense of hearing in fish. Nature (Lond) 141:8–11 von Humboldt A: Voyage aux régions équinoxiales du Nouveau Continent fait en 1799 et 1800 par A. de Humboldt et A. Bonpland: l'Orénoque. Gheerbrant A (ed), Club des Libraires de France, Paris 1961

Walker MM (1984) Learned magnetic field discrimination in yellowfin tuna, *Thunnus albacares*. J Comp Physiol 155:673–679

Watanabe A, Takeda K (1963) The change of discharge frequency by a.c. stimulus in a weak electric fish. J Exp Biol 40:57–66

Waxman SG, Pappas GD, Bennett MVL (1972) Morphological correlates of functional differentiation of nodes of Ranvier along single fibers in the neurogenic electric organ of the knife fish, *Sternarchus*. J Cell Biol 53:210–224

Wells KD (1977) The social behaviour of anuran amphibians. Anim Behav 25:666–693

Westby GWM (1974) Assessment of the signal value of certain discharge patterns in the electric fish, *Gymnotus carapo*, by means of playback. J Comp Physiol 92:327–341

Westby GWM (1975a) Comparative studies of the aggressive behaviour of two gymnotid electric fish (*Gymnotus carapo* and *Hypopomus artedi*). Anim Behav 23:192–213

Westby GWM (1975b) Further analysis of the individual discharge characteristics predicting social dominance in the electric fish, *Gymnotus carapo*. Anim Behav 23:249–260

Westby GWM (1975c) Has the latency dependent response of *Gymnotus carapo* to discharge-triggered stimuli a bearing on electric fish communication? J Comp Physiol 96:307–341

Westby GWM (1979) Electrical communication and jamming avoidance between resting *Gymnotus carapo*. Behav Ecol Sociobiol 4: 381–393

Westby GWM (1984) Simple computer model accounts for observed individual and sex differences in electric fish signals. Anim Behav 32:1254–1256

Westby GWM (1988) The ecology, discharge diversity and predatory behaviour of gymnotiforme electric fish in the coastal streams of French Guiana. Behav Ecol Sociobiol 22:341–354

Westby GWM, Box HO (1970) Prediction of dominance in social groups of the electric fish, *Gymnotus carapo*. Psychon Sci 21:181–183

Westby GWM, Kirschbaum F (1977) Emergence and development of the electric organ discharge in the mormyrid fish, *Pollimyrus isidori*. I. The larval discharge. J Comp Physiol 122:251–271

Westby GWM, Kirschbaum F (1978) Emergence and development of the electric organ discharge in the mormyrid fish, *Pollimyrus isidori*. II. Replacement of the larval by the adult discharge. J Comp Physiol 127:45–59

Westby GWM, Kirschbaum F (1981) Sex differences in the electric organ discharge of *Eigenmannia virescens* and the effect of gonadal maturation. In: Szabo T, Czeh G (eds) Sensory physiology of aquatic lower vertebrates. Adv Physiol Sci, Pergamon Press / Akademiai Kiado, Budapest, 31:179–194

Westby GWM, Kirschbaum F (1982) Sex differences in the waveform of the pulse-type electric fish, *Pollimyrus isidori* (Mormyridae). J Comp Physiol 145:399–403

Westby GWM, Shepperd DK (1986) Waveform recognition in weakly-electric fish. SERC Bull 3:6–7

Wickler W (1980) Vocal dueting and the pair bond. I. Coyness and partner commitment. A hypothesis. Z Tierpsychol 52:201–209

Wilson EO (1975) Sociobiology. The new synthesis. Harvard Univ Press, Cambridge (Massachusetts)

Wohlfahrt TA (1939) Untersuchungen über das Tonunterscheidungsvermögen der Elritze (*Phoxinus laevis* Agass.). Z Vgl Physiol 26:570–604

Wu CH (1984) Electric fish and the discovery of animal electricity. Am Sci 72:598–607

Wyman (1965) Probabilistic characterization of simultaneous nerve impulse sequences controlling dipteran flight. Biophys J 5:447–471

Zakon HH (1986) The electroreceptive periphery. In: Bullock TH, Heiligenberg W (eds) Electroreception. Wiley, NY, pp 103–156

Zakon HH (1988) The electroreceptors: diversity in structure and function. In: Atema J, Fay RR, Popper AN, Tavolga WN (eds) Sensory biology of aquatic animals. Springer, Berlin Heidelberg New York Tokyo, pp 813–850

Zelick R, Narins PM (1985) Characterization of the advertisement call oscillator in the frog *Eleutherodactylus coqui*. J Comp Physiol 156:223–229

Zimmermann H (1985) Die elektrischen Fische und die Neurobiologie: über die Bedeutung einer naturgeschichtlichen Kuriosität für die Entwicklung einer Wissenschaft. Funkt Biol Med 4:156–172

Zipser B, Bennett MVL (1976a) Responses of cells of the posterior lateral line lobe to activation of electroreceptors in a mormyrid fish. J Neurophysiol 39:693–712

Zipser B, Bennett MVL (1976b) Interaction of electrosensory and electromotor signals in the lateral line lobe of a mormyrid fish. J Neurophysiol 39:713–721

Zwicker E (1982) Psychoakustik. Springer, Berlin Heidelberg New York

Systematic Index

Adontosternarchus 12, 13, 83
Adontosternarchus balaenops 14
Amia 17
Apteronotidae 12–14, 27, 42, 46–48, 61, 75, 80–83, 100, 107, 210
Apteronotus 12, 21, 39, 80, 82, 83, 108, 135, 200–202
Apteronotus albifrons 51, 56, 80, 82–84, 100–102, 184, 187
Apteronotus anas 80, 83
Apteronotus bonaparti 80, 83
Apteronotus hasemani 80, 83
Apteronotus leptorhynchus 53, 54, 101, 102, 136, 187, 196, 200
Apteronotus rostratus 107
Archolaemus 12
Ariidae 15
Astroscopus 39, 40, 42, 45

Batoidea 38
bichir, see *Polypterus*
Boulengeromyrus 8
Brienomyrus 8
Brienomyrus brachyistius 6, 63, 88, 94, 122, 167
Brienomyrus niger 6, 57, 86–90, 92–95, 104, 118, 124, 131, 139–146, 158–160, 163–169

Campylomormyrus 8
Campylomormyrus numenius 6, 7
Campylomormyrus tamandua 6, 7
cartilaginous fishes, see Chondrichthyes
catfish 15, 19
   electric, see *Malapterurus electricus*
   ictalurid 17
characiforms 4, 9, 13
Chondrichthyes 17, 18, 38, 58
Chondrostei 18
cladistia 18

Dipnoi 18
Distocyclus 12, 61, 75, 76, 79, 83
Distocyclus conirostris 75, 76
Distocyclus goachira 61, 79

Eigenmannia 12–14, 32, 36–38, 42, 45, 47, 48, 54, 56, 60, 62, 66, 69, 75–79, 82, 83, 101, 102, 108, 135, 181, 182, 185–188, 195, 200–203, 208, 210, 213–215
Eigenmannia lineata 77, 79, 83, 101, 102, 136, 188–191, 193–195, 204, 207, 210–212, 214
Eigenmannia macrops 76, 79, 82, 83, 101, 197, 204
Eigenmannia virescens 10, 14, 76, 79, 83, 102, 103, 135, 136, 184, 196, 197
Electrophoridae 12, 14, 15, 27, 70
Electrophorus electricus 12, 14, 15, 38–40, 42, 44–46, 70–72, 74, 75, 85, 99, 111, 178

gar 17
Genyomyrus 8
Gnathonemus 8, 105
Gnathonemus petersii 6, 7, 21, 26, 31, 39, 54, 60, 62–65, 86, 88–91, 94, 95, 111–125, 131, 137–158
Gnathonemus senegalensis, see *Marcusenius senegalensis*
Gymnarchidae 5, 8, 10, 23, 27, 61, 62, 97
Gymnarchus niloticus 8, 10, 23, 25, 27, 41, 45, 50, 54, 61, 62, 69, 70, 97, 106, 131, 132, 194
Gymnorhamphichthys 12, 13
Gymnorhamphichthys hypostomus 14, 98–100, 135, 184
Gymnotidae 12, 14, 15, 27, 70
Gymnotus 12, 24, 28, 39, 45, 46
Gymnotus anguillaris 70
Gymnotus carapo 10, 14, 15, 70, 71, 72–74, 97–100, 106, 107, 132–134, 178, 179, 181, 182, 200

Heteromormyrus 8
Hiodontidae 5
Hippopotamyrus 8
Hippopotamyrus harringtoni 6, 7
Hydrocynus forskalii 9
Hyperopisus 8
Hypopomidae 12–14, 27, 46, 70

235

Hypopomus 12, 21, 32, 45, 48, 71, 72, 74, 85, 98–100, 106, 107, 181, 182
Hypopomus artedi 132, 135
Hypopomus beebei 73, 178
Hypopomus brevirostris 134
Hypopomus occidentalis 14, 66, 73, 134, 135, 179, 181
Hypopygus 12, 72, 98, 180–182

Isichthys 8
Ivindomyrus 8

jawed fishes 18

Kryptopterus 29

lamprey 18
Lepomis cyanellus 178
lungfish, see Dipnoi

Malapteruridae 15
Malapterurus electricus 38-40, 48, 49, 55, 84–86, 103, 108, 109
Marcusenius 8, 124
Marcusenius cyprinoides 6, 7, 88, 92, 94, 95, 158
Marcusenius greshoffi 6, 7
Marcusenius macrolepidotus 86, 87, 177
Marcusenius senegalensis 111
Mochokidae 15
Mormyridae 5, 7, 8, 27, 33, 45–50, 60, 62, 63, 71, 72, 86, 104, 112, 125, 130, 180
Mormyrops 8
Mormyrops deliciosus 6, 7, 62, 86, 88, 94, 111
Mormyrops zanclirostris 111
Mormyrus 8
Mormyrus caballus 21
Mormyrus kannume 9
Mormyrus rume 6, 7, 40, 46, 62, 63, 70, 86, 88, 92, 94, 95, 113–116, 118–122, 124

Narcine 108
Neopterygii 17–19
Notopteridae 5, 6
Notopteroidei 5
Notopterus afer, see *Papyrocranus afer*

Oedemognathus 12
Oedemognathus exodon 80–83
Orthosternarchus 12
Ostariophysi 4, 18–20, 107, 176

Osteichthyes 18
  phylogeny 4, 18
Osteoglossidae 5
Osteoglossiformes 5, 6, 20
Osteoglossomorpha 4, 18–20

paddlefish 18
Pantodontidae 5
Papyrocranus afer 5, 6, 20
Paramormyrops 8
Parupygus 12
Petrocephalus 8
Petrocephalus bovei 6, 7, 63, 66, 86, 88, 94, 95, 124, 158–170
Plotosidae 15
Plotosus 19, 23
Pollimyrus 8
Pollimyrus isidori 6, 9, 47, 54, 62–69, 73, 86–96, 117, 121–131, 157–160, 162, 164–177, 180
Polypterus 18, 109
Porotergus 12
Porotergus gymnotus 80, 81, 83
Prochilodus scrofa 13
Pseudocetopsis 103

Raja 39, 41
Rhabdolichops 12, 13, 43, 54, 76, 83
Rhamphichthyidae 12–14, 27, 70
Rhamphichthys 12, 13, 71, 72, 74, 98, 180, 182, 183, 200
Rhamphichthys rostratus 182

Siluriformes 4, 15, 20, 84, 103, 108
stargazer, see *Astroscopus*
Steatogenys 12, 13, 98, 100, 133
Steatogenys elegans 14, 71, 72, 98, 100, 178
Steatogenys duidae 100
Sternarchella 12, 61, 80–83
Sternarchella schotti 80, 81, 83
Sternarchidae, see Apteronotidae
Sternarchogiton 12, 80, 81
Sternarchogiton naterreri 80, 83, 101
Sternarchorhamphus 12, 82, 198–200
Sternarchorhamphus macrostomus 80, 81, 83, 184
Sternarchorhynchus 12, 199, 200
Sternarchorhynchus curvirostris 80, 81, 83
Sternarchorhynchus mormyrus 80, 81, 83
Sternarchorhynchus oxyrhynchus 83
Sternarchus, see *Apteronotus*
Sternopygidae 12–14, 26, 27, 30, 43, 46, 47, 70, 75, 76, 79, 82, 100, 107, 187, 215

Sternopygus 12, 13, 47, 78, 101, 184, 187, 188, 197, 200, 201
Sternopygus dariensis 102
Sternopygus macrurus 10, 76, 83, 102, 136, 184, 187, 200, 215
Stomatorhinus 8
sturgeon 18
Synodontis 15, 55, 84, 103

Teleostei 18
  phylogeny 4, 18

Ubidia 12

Xenomystinae 5
Xenomystus nigri 5, 6, 20, 178
Xenopus 176

Subject Index

acoustic sensitivity 4, 104, 107, 130, 131
aggression 112–122, 126, 131–133, 135, 136, 139, 140–151, 153–157, 178, 179, 184, 196, 215
ampullae of Lorenzini 17, 28
androgen hormone 47, 65, 102
anterior exterolateral toral nucleus 33, 34
autocorrelation 89, 90, 93, 95, 148, 149, 151, 154–156, 168

B unit, see burst duration coder
beat 61, 192, 194, 196, 198, 200, 201, 209
"best" frequency, see stimulus filtering
burst duration coder 26, 27, 32, 33, 54

canal organ 17, 21–23
caudal cerebellar lobe (LC) 34
central expectation 35
cerebellum 7, 22, 33–35, 177
circadian rhythm 13, 59, 61, 86, 93, 97–101, 104, 106, 108, 125, 167, 170
cladistic phylogeny 4, 5, 18, 19
club ending 50
command nucleus, medullary 50
communication range 52, 57, 58, 66
conditioned responses 105, 170–177, 202, 203, 210, 211, 215
corollary discharge 35, 36, 50, 53, 180
courtship 89, 118, 125–128, 130, 131, 134–136, 184
cranial nerves 19, 22
cross-correlation 151, 155–157

desmosomes (tight junctions) 24, 25
difference limen, see discrimination limen
digital synthesis of electric organ discharges 136, 196, 197, 203, 205–209, 211–215, 217
dipole field 51–54, 57, 58
discrimination limen 170–177, 213

echo response, see Preferred Latency Response
echolocation 41
efference copy, see corollary discharge

ELa, see anterior exterolateral toral nucleus
electrocyte 42–46, 48, 74, 78, 82, 84
electromotoneuron 42, 48–51, 75
electroreception, phylogeny 4
electroreceptors 23, 26, 33, 48, 53, 217
 ampullary 4, 15, 17–19, 21, 23–25, 28–30, 33–37, 40, 54, 196
 tuberous 4, 19–21, 23–26, 32, 36, 40, 53, 54, 73, 76, 102, 103, 192, 200, 201
ELL, electrosensory lobe of the lateral line 27, 30, 33–37, 50
ELp, see posterior exterolateral toral nucleus
eminentia granularis
 posterior (EGp) 34, 37, 50
 medialis (EGm) 37
end bud 18, 19
EOCD, see corollary discharge
EOD, electric organ discharge 59
EOD sound response 104, 107
equipotentials, see isopotentials
estrogen hormone 102
exafference 33

Fourier amplitude spectrum (of EOD) 32, 60, 65, 68, 71–73, 76, 77, 79–81, 83, 154, 155, 157, 201, 203

gap junction, see synapse (electrotonically transmitting)
Gemminger's bones 7
gymnarchomast 23, 25, 27, 28

habituation 187, 205
hair cell 18, 19, 26
harmonic content 60–62, 75–77, 80, 82, 83, 146, 203, 208, 209, 213, 215
hearing, see acoustic sensitivity
Hunter's organ 45

ichthyological provinces
 Africa 8, 9, 11, 12, 15
 South America 11–13
impedance 33, 53–55, 105, 108
 resistive 33, 53–55, 105, 106, 195
 capacitive 33, 54, 61, 105

inferior colliculus 36
inferior lobe 37
inferior olivary nucleus 37
information theory 142–146
isopotentials 43, 51–53

jamming avoidance response 36, 38, 62, 76, 107, 108, 134, 135, 180–200, 203–208, 210, 213, 215
JND, see discrimination limen

kinocilium 18, 24
Knollenorgan 21, 25–30, 32–36, 53, 68, 103, 125

larval electric organ 42, 47
lateral inhibition 36
lateral line nerve 4, 33
 anterior 19, 20, 22, 23, 36
 posterior 19, 20, 22, 23, 31
lateral line system 17, 22
 receptors 17, 200, 217
lateral toral nucleus 34
lemniscus lateralis 33, 36, 37

M unit, see pulse marker unit
magnetic field 1, 54, 105, 106
main organ (of electric eel) 45
medulla 36
medullary relay nucleus 37, 38, 48–51, 75
mesencephalon 33, 34, 50
metencephalon 36
microtubules 18
microvilli 18, 24, 26, 27
monoamines 38, 102
mormyromast 21, 25–36, 53, 95, 125

nELL, see nucleus of the electrosensory lateral line lobe
neuromast 17, 23, 33
nucleus of the electrosensory lateral line lobe (nELL) 34, 35
nucleus electrosensorius 38
nucleus praeeminentialis 36
 dorsal 37

optic tectum 34, 37

P unit, see probability coder
pacemaker (of electric organ) 37, 38, 44, 48–50, 102, 107, 182, 185
phase coder (T unit) 26, 27, 32, 33, 36–38, 54, 192, 195, 196, 209, 212
phase coupling (active) 136, 181, 184, 198–200, 203
phase sensitivity 180–183, 195, 196, 198
posterior exterolateral toral nucleus 34

precommand nucleus 50
preferred latency avoidance (PLA) 121–125, 180
preferred latency response (PLR) 30, 32, 105, 120–125, 135, 138, 149, 172, 180, 200
preeminential nucleus, see nucleus praeeminentialis
prepacemaker nucleus · 37, 38, 48
presynaptic ribbon 18
probability coder (P unit) 26, 27, 32, 33, 36, 37, 54, 192, 195, 196, 209
pulse marker unit (M unit) 26, 27, 32, 54
PSP (postsynaptic potential) 45, 49, 50, 74, 75

Q_{10} 63, 78, 92, 101

reactance, see impedance
reafference 33, 35, 36, 50, 53
reticular formation 34, 50
rhombencephalon 33

Sachs' organ 45, 75
schooling 52, 121, 125, 170, 199
sexual dimorphism 61–65, 76–78, 102, 103, 122, 123, 136, 203, 209, 210, 213–215, 217
skin resistivity 24, 26
sound production 125, 129–131
spawning 89, 118, 125–131, 134, 136, 184
spontaneous preference 92, 157–170, 203, 213–215
stimulus filtering 32, 68, 69, 73, 102, 179, 200–202, 209, 213, 215, 217
stratum spinosum 23
symplesiomorphy 4
synapomorphy 4
synapse 26
 chemical 26
 electrotonically transmitting 26, 33, 34, 43, 48

T unit, see phase coder
tectum 38
telencephalon 37
tonofilament 24, 25
torus semicircularis 33, 36–38
transmitter 29, 38, 41, 42, 102
tuning (of receptor), see stimulus filtering

valvula 34

water conductivity 41, 47, 52, 57, 58, 63, 66–69, 73, 74, 78, 84, 213
water resistivity, see water conductivity
Weber's law 177, 204